P. Margaretha

Preparative Organic Photochemistry

P. Margaretha

Preparative Organic Photochemistry

With 9 Figures

AKADEMIE-VERLAG · BERLIN
1982

Die Originalausgabe erscheint im Springer-Verlag
Berlin · Heidelberg · New York

Vertrieb ausschließlich für die DDR und die sozialistischen Länder

Erschienen im Akademie-Verlag, DDR-1086 Berlin, Leipziger Straße 3–4
Lizenznummer: 202 · 100/547/82
Gesamtherstellung: VEB Druckerei „Thomas Müntzer", 5820 Bad Langensalza
Umschlaggestaltung: Karl Salzbrunn
Bestellnummer: 763 157 1 (6728) · LSV 1275
Printed in GDR
DDR 48,— M

Managing Editor:

Dr. *Friedrich L. Boschke*
Springer-Verlag, Postfach 105280, D-6900 Heidelberg 1

Editorial Board:

Preface

It is now more than 13 years ago that the second edition of A. Schönberg's "Preparative Organic Photochemistry" appeared, a work representing a thorough compilation of photochemical reactions of preparative concern to the organic chemist.

Research in organic photochemistry over this period has not only produced a vast accumulation of factual knowledge, but the understanding of *how* these reactions proceed has also dramatically increased. The high rate of development in synthetic applications of photochemical reactions has — at least in part — only become possible due to this theoretical and mechanistic know-how. Organic photochemistry is today a multidisciplinary field itself.

It is not the purpose of this book to be a supplement to the earlier work bearing the same title. The intent is to demonstrate the usefulness of organic photochemical reactions as powerful tools in organic synthesis. In contrast to the preceding monograph, an exhaustive compilation of all preparative applications of photochemical reactions has become virtually impossible in this vigorously expanding field. I have therefore tried to select the (subjectively) most representative examples of synthetic utile reactions with a strong preference for more recent publications and reviews. In this context the literature was covered up to June 1981.

Hamburg, November 1981 P. Margaretha

Table of Contents

1 Introduction

1.1 Photochemistry

Chemical reactions can be differentiated fundamentally by describing the electronic configuration of the reactive species itself. On the one side there are species with unoccupied antibonding orbitals (ground-state molecules, radical cations), on the other side we deal with species with occupied antibonding orbitals (radical anions, excited molecules). While radical cations and radical anions are usually generated by electron transfer from or to the ground state molecule, the excited species is formed by promotion of an electron from a bonding or a nonbonding orbital to an antibonding orbital (Fig. 1).

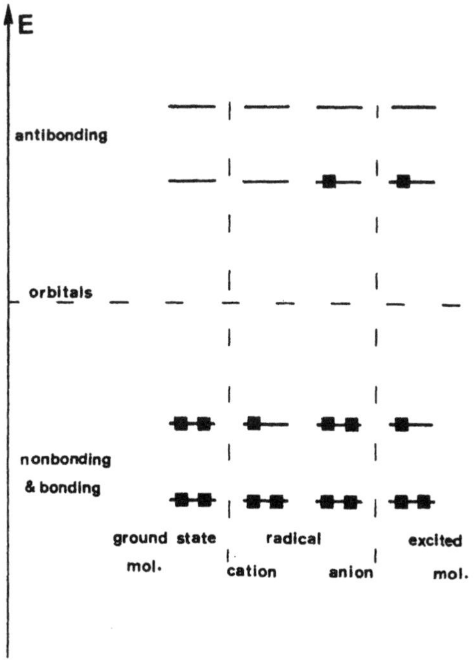

Fig. 1. Electronic configuration of intermediates obtained by electron transfer or by excitation

Excitation of organic molecules is usually achieved with visible or ultraviolet light (λ = 750–200 nm) corresponding to a difference in energies between the excited state of a molecule and its ground state of about 40–145 kcal/mole. As the excited state is created directly by absorption of a light quantum ground state reactions from higher vibrational levels, e.g. bond breaking, are avoided although the amount of energy would be largely sufficient for such a process to occur. The excited molecule posesses the intrinsic tendency of either liberating this excess energy or of undergoing chemical reactions. Photochemistry therefore deals with chemical changes undergone by such excited molecules while the purely physical processes of setting this excess energy free give back the molecule in its ground state. Radiation chemistry has to do with the effects of more energetic radiation (λ < 150 nm) and although related to photochemistry, is traditionally dealt with separately.

Photochemistry as topic is covered in several introductory textbooks [101–107]. The annual literature is surveyed in a specialist periodical report [108]. Two series of monographs have to do with selected chapters from organic photochemistry [109] or photochemistry in general [110]. A series on molecular rearrangements also covers photochemical reactions [111].

1.2 Generation of Excited Molecules

The excited states of a molecule are characterized by their multiplicity (usually singlet or triplet states) and by the difference in energy between the excited state itself and the ground state (Fig. 2).

While in the thermal or electrochemical activation of a molecule the energy is transfered in a continuous way, photochemical activation is insofar unique as it occurs by absorption of a photon of well defined energy. Although there are chemical ways of generating excited states [112], under normal conditions a molecule can only be excited by absorption of light of the wavelength corresponding to the energy difference between the excited state and the ground state. This is known as Planck's law (1.1).

$$E_n - E_0 = h\nu_n = \frac{hc}{\lambda_n} \qquad (1.1)$$

where h is Planck's constant, ν is the frequency at which absorption occurs and c is the speed of light. The position of the absorption band is more often expressed by the wavelength λ in nanometers (1 nm = 10^{-9} m). The

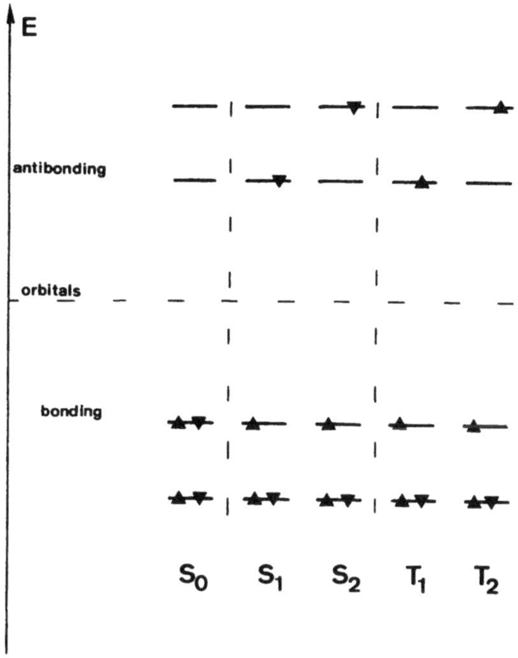

Fig. 2. Electronic configuration of the first two excited singlet and triplet states

conversion of energy (kcal/mole) to the wavelength λ (nm) is given by the following expression (1.2).

$$\Delta E\ (\text{kcal/mole}) = \frac{2.86 \times 10^4}{\lambda\ (\text{nm})} \tag{1.2}$$

Continuous energy transfer implies, that *A* giving *B* at a temperature T_1 (or potential E_1) and giving *C* at a temperature T_2 (or potential E_2), that *B* will also convert to *C* under these "higher energy" conditions. This is so for all thermal or electrochemical reactions. In photochemical reactions *A* gives *B* (an excited state) with light of wavelength λ_1 and *C* (a higher excited state [113,114]) with light of shorter wavelength λ_2; but now *B* will *not* give *C* with λ_2 as this wavelength does not correspond to the difference in energies between the excited states *B* and *C* (Fig. 3).

Before irradiating a compound one has therefore to know its absorption spectrum. A typical absorption spectrum is shown in Fig. 4, affording the following information:

a) the energy(ies) of the excited state(s) can be deduced from the wavelength where the absorption curve ends towards longer wavelengths
b) the degree of allowedness of a transition is reflected by the intensity of the band.

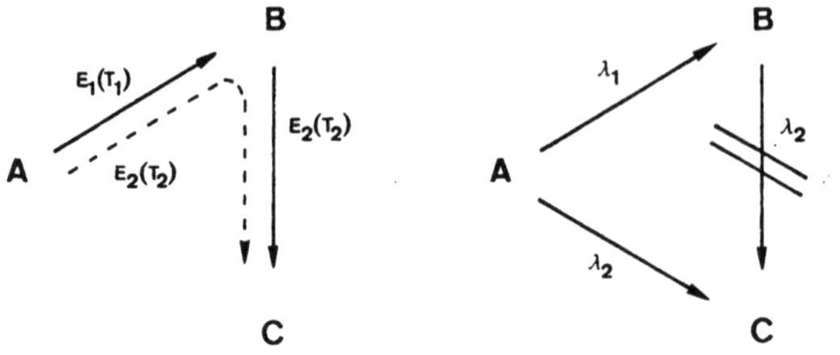

Fig. 3. Continuous energy transfer versus discontinuous energy transfer

This intensity is expressed by the molar absorption coefficient ε which can be calculated from the (measured) absorbance A, (A = log I_0/I) via the well known equation of Lambert & Beer (1.3), wherein c is the concentration (mole/l) and d is the optical path length of the cell (in cm).

$$A = \varepsilon \cdot c \cdot d \tag{1.3}$$

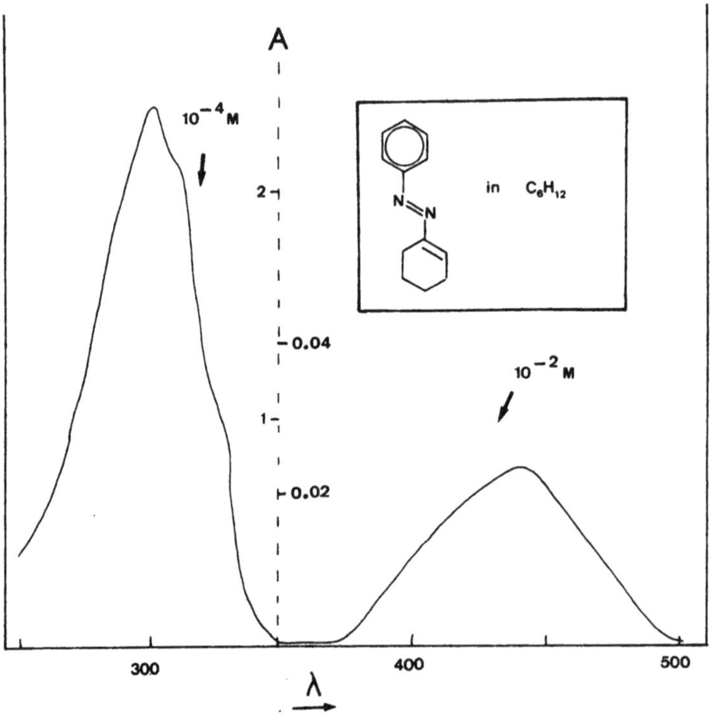

Fig. 4. UV.-absorption spectrum of phenylazocyclohexene in cyclohexane as solvent

While for thermal reactions one usually does not correlate the energy input with the amount of product formed, electrochemists and photochemists are certainly more "energy-minded". The first ones use the current yield to define the amount of product formed per electrons consumed. The latter ones use the so called quantum yield Φ which is defined as the ratio of number of molecules undergoing a particular process from an excited state over moles of photons absorbed by the system, or in other words, the ratio of the rate constant for the process defined over the sum of all rate constants for all possible processes from this excited state (1.4). Thus, if for every photon absorbed, a molecule undergoes only one chemical process, the quantum yield for this process is unity; if other processes compete it will be less than unity.

$$\Phi_x = \frac{k_x}{k_a + k_b + \ldots + k_x} \qquad (1.4)$$

For the conversion of the molecule *AB* to the Product *P* the quantum yield Φ for the formation of *P* corresponds to the number of molecules *P* formed over the number of light quanta absorbed by *AB*. For such a reaction one should not correlate quantum yield and chemical yield. In (1.5) an example of a reaction is given where the quantum yield of product formation is very low although the chemical yield is high. In (1.6) the chemical yield is zero although the photochemical step itself is highly efficient. In this latter case the thermolabile intermediate *AB'* can obviously be trapped before reconverting to starting material, or the formation of *P* can be favoured by running the reaction at low temperature.

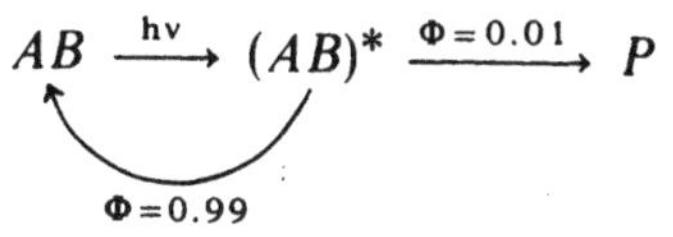

(1.5)

q.y. = 0.01
c.y. = 100 %

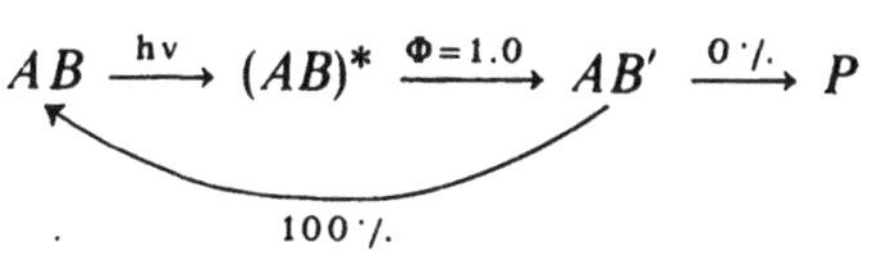

(1.6)

q.y. = 1.0
c.y. = 0 %

1.3 Excited Molecules — Physical Processes

By absorption of light a molecule is promoted to a higher electronic state. The monomolecular physical processes for the dissipation of the excess energy are outlined in Fig. 5 in a so called Jablonski diagramm. In principle one has to differentiate between radiative and non-radiative deactivation on the one side and on the other side one has to consider if the multiplicity of the system is conserved or not. Radiative deactivation, i.e. deactivation accompanied by emission of light, is termed
fluorescence if the transition occurs with spin conservation and
phosphorescence, if spin inversion occurs.

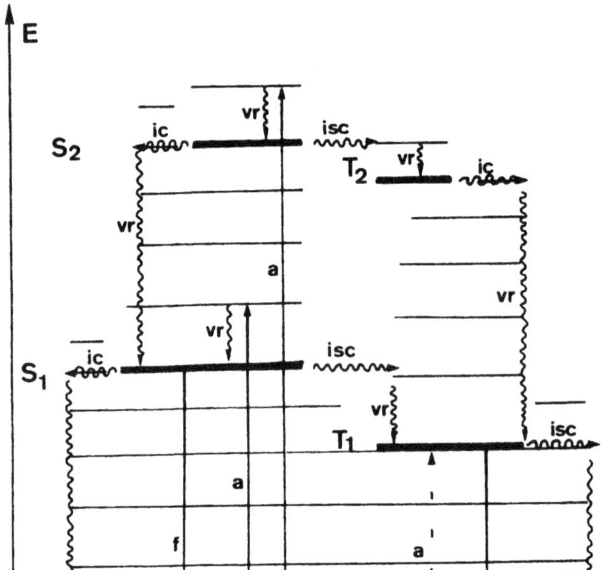

Fig. 5. Jablonski diagram

a) absorption *vr*) vibrational relaxation

deactivation → non radiative → *ic*) internal conversion *; *isc*) intersystem crossing **

deactivation → radiative → *f*) fluorescence *; *p*) phosphorescence **

* $\Delta S = 0$ ** $\Delta S \neq 0$

The radiationless transition between two states of same spin is called *internal conversion*, the one occuring with inversion of spin being termed *intersystem crossing*. In both processes the excess energy is liberated as heat. All these transitions between different electronic states are customarily preceded by vibrational relaxation, i.e. the deactivation from a higher vibronic level to the v_0-level of the same electronic state (Fig. 5).

As shown in Fig. 6, there is a correlation between absorption spectrum and emission spectrum. Taking into consideration the Franck-Condon principle, which states that there is no motion of the atoms during an electronic transition, one has to differentiate between the two following possibilities:
in the one the geometry of the excited state is similar to the one of the ground state (same interatomic distances),
in the other one the geometries are different (usually the distances between the atoms become larger in the excited state).

In the first case the so called 0—0 transition will be the most probable one, both for absorption and emission, and therefore the most intense

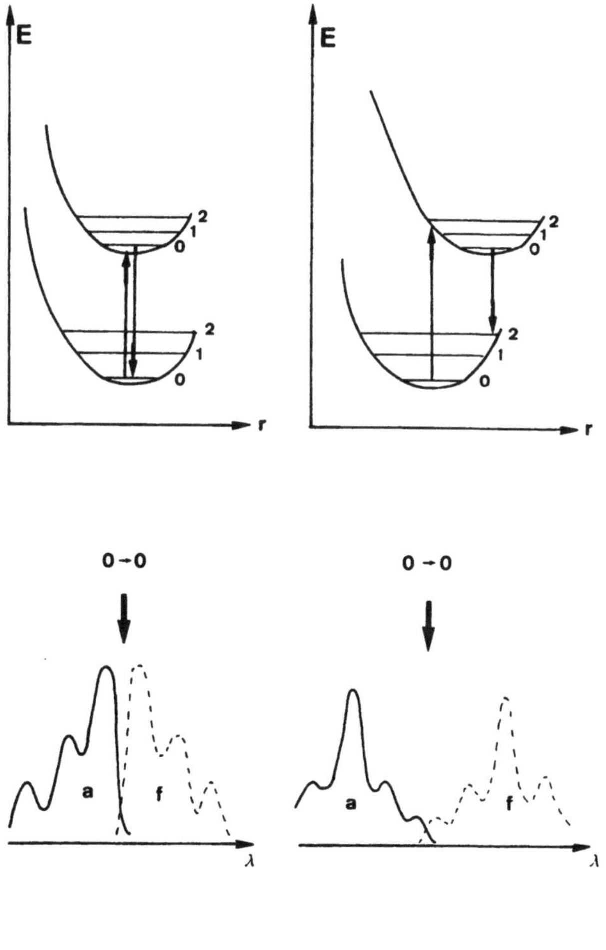

Fig. 6. Correlation between absorption and emission (fluorescence) spectrum

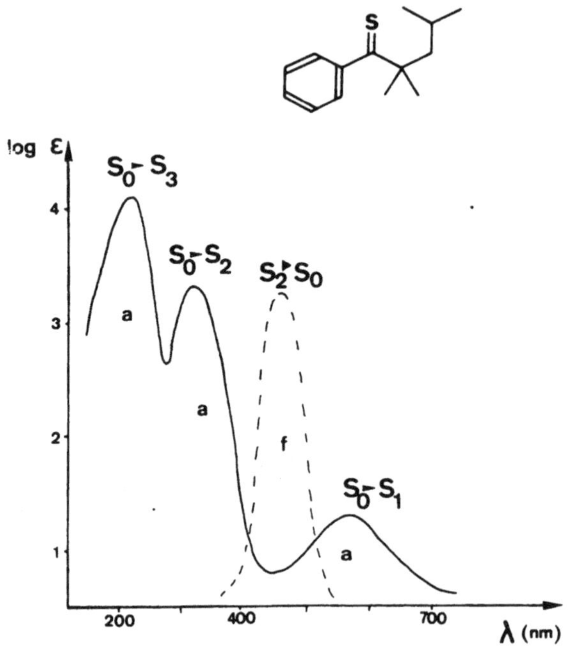

Fig. 7. UV.-absorption spectrum (in C_6H_{12} at room temperature) and fluorescence spectrum (in ether/pentane at 77 K) of an aryl alkyl thioketone

in the spectra. In the second case the most probable transitions will be e.g. 0–2′ and 0′–2. In both instances a mirror image relationship between absorption spectrum and emission spectrum holds. This does not only allow to determine the energy difference between excited state and ground state but it also makes it possible to determine the emitting state itself. As demonstrated in Fig. 7 the emitting (fluorescing) state for the alkyl aryl thioketone is S_2 and not S_1 as evidenced by the absorption — emission mirror image relation.

Besides the excited molecule can interact physically with a second molecule, i.e. undergo *bimolecular processes*. These are either energy transfer (1.7) or exciplex formation (1.8) depending on the relative excitation energies of the molecule to be studied and its partner.

$$A^* + B \rightarrow B^* + A \quad (1.7)$$

$$A^* + B \rightarrow (A^* - B) \quad (1.8)$$

Energy transfer can occur if the difference in energies between A^* and A is larger than the one between B and B^*. Two mechanisms for such an energy transfer are common: a so called physical mechanism were the two partners do not get in contact with each other and the so termed chemical mechanism where the two partners approach each other sufficiently to form an encounter complex. An important feature in energy transfer is, that the spin selection rules are less severe than in monomolecular transitions so that energy transfer from an excited molecule in its triplet state (the donor *D*) to a partner in its singlet ground state (the acceptor *A*) is possible, leading to a direct population of the triplet state of *A* (1.9). The overall spin of the system is hereby conserved.

$$^3D_1 + {}^1A_0 \xrightarrow{ET} {}^1D_0 + {}^3A_1 \qquad (1.9)$$

This process — also called *sensibilisation* — is of importance in preparative organic photochemistry [115] whenever a rate constant of intersystem crossing $k_{S_1 \to T_1}$ for a given molecule is small or if direct irradiation of this molecule is inconvenient. This principle is illustrated in Figure 8 for benzophenone as donor and a diene as acceptor, the diene then isomerizing from the triplet state.

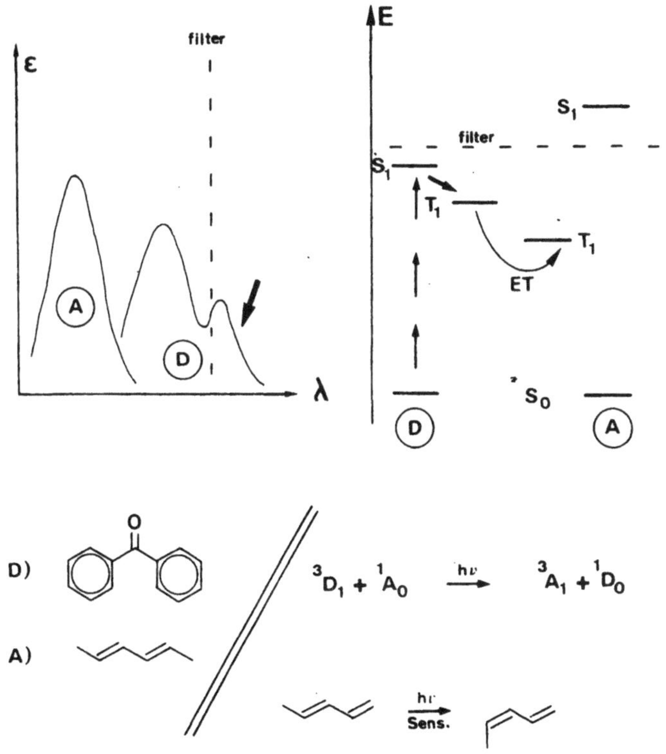

Fig. 8. Benzophenone-sensitized isomerization of E-pentadiene. UV.-absorption spectra and energy diagramm

Exciplexes[116,117] can be formed if the excitation energy $B \rightarrow B^*$ is higher than the one for $A \rightarrow A^*$ in (1.8). Such an *excited* com*plex* is associative in the excited state only, the corresponding ground state complex between A and B being dissociative (Fig. 9). Such exciplexes are important intermediates in e.g. cycloaddition reactions as precursors of diradicals[118] which are themselves precursors of the cyclized photoproducts.

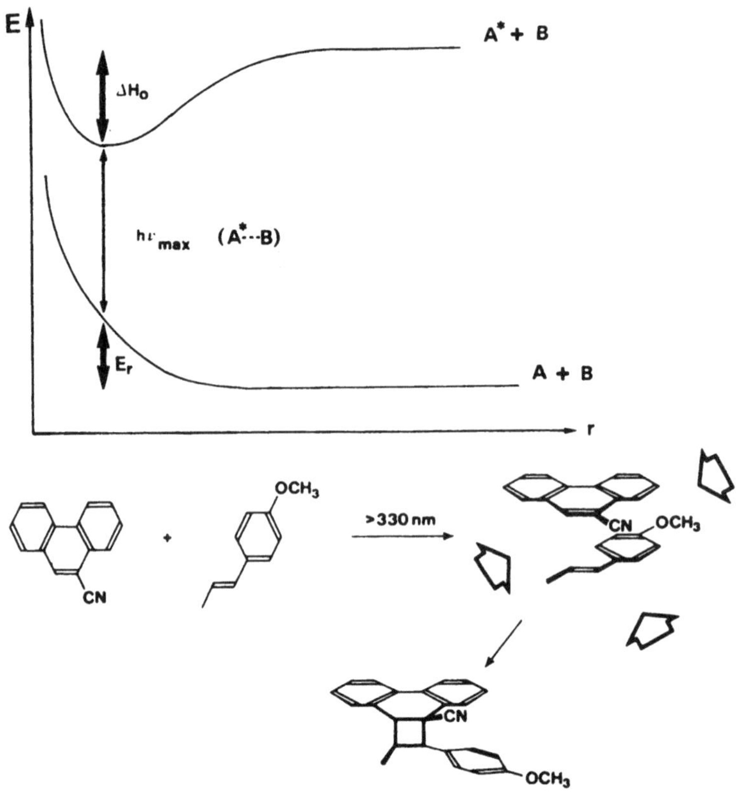

Fig. 9. Exciplex formation between excited 9-cyanophenanthrene and p. methoxystyrene. E_r is the ground-state repulsion energy; $h\nu_{max}$ is the emission maximum of the exciplex and H_0 the stabilization energy of the exciplex

1.4 The Behaviour of Excited Molecules — Chemical Processes

Many different examples of preparative useful photochemical reactions are going to be discussed in the following chapters. Nevertheless the mechanistic possibilities for an excited molecule to undergo chemical reactions are rather limited. One way of classifying photochemical reactions consists in describing the initial photochemical event. Another approach of classifying photochemical reactions by their topicity [119)] has been suggested recently.

For a molecule *AB* the primary monomolecular photochemical processes which can occur are:

a) *heterolytic bond cleavage* (1.10), a rather uncommon process for organic molecules;
b) *homolytic bond cleavage* to a pair of radicals or a diradical (1.11);
c) *photoextrusion of a small molecule* as N_2, CO or CO_2 (1.12), usually a two step reaction wherein the first step is again a homolytic bond breaking;
d) *rearrangement* to a new isomer *AB'* (1.13).

This isomerization can lead to a molecule with different constitution, configuration or conformation, or to a valence isomer.

$$AB \xrightarrow{h\nu} (AB)^* \rightarrow A^+ + B^- \tag{1.10}$$

$$AB \xrightarrow{h\nu} (AB)^* \rightarrow A^{\cdot} + B^{\cdot} \tag{1.11}$$

$$AB \xrightarrow{h\nu} (AB)^* \rightarrow A^{\cdot} + B^{\cdot} \rightarrow A^{\cdot} + C^{\cdot} + X \tag{1.12}$$

$$AB \xrightarrow{h\nu} (AB)^* \rightarrow AB' \tag{1.13}$$

In the presence of a proper second molecule bimolecular photochemical processes occur. Obviously such reactions can also occur in an intramolecular fashion in bifunctional molecules containing both reactive centres. These reactions are:

a) *hydrogen abstraction* by the excited molecule if the second molecule (or reactive centre) is a hydrogen donor *RH* (1.14);
b) *photodimerisation* (1.15);
c) *photoaddition* or photocycloaddition (1.16);
d) *electron transfer* (1.17), if no bonding takes place between the reactants (or reactive centres).

$$AB \xrightarrow{h\nu} (AB)^* \xrightarrow{RH} ABH^{\cdot} + R^{\cdot} \tag{1.14}$$

$$AB \xrightarrow{h\nu} (AB)^* \xrightarrow{AB} (AB)_2 \tag{1.15}$$

$$AB \xrightarrow{h\nu} (AB)^* \xrightarrow{C} ABC \tag{1.16}$$

$$AB \xrightarrow{h\nu} (AB)^* \xrightarrow{C} AB^+ + C^- \quad \text{or} \quad AB^- + C^+ \tag{1.17}$$

There are books [120,121] and review articles [122–124] dealing with preparative organic photochemistry in general, each author using his own system to group and classify the reactions described. Photochemical preparations *per se* have also been gathered [125]. Not immediate preparative utilizations as photo-removable protecting groups have also been discussed [126].

In the next chapters preparative organic photochemical reactions will be discussed in the sequence given above, i.e. (1.11) → (1.17), the following system being adopted for the subdivisions of each chapter: for photocleavage reactions the atomic weight of the bond forming atoms will increase, as does the molecular weight of the chromophore for photorearrangements and photoaddition reactions. The choice of the reactions presented as examples is necessarily a subjective one. One main concern has been to select as recent reports and reviews from the literature as possible.

2 Photochemical Cleavage Reactions in Synthetic Organic Chemistry

2.1 Photofragmentation Reactions

This chapter has to do with reactions wherein the photochemical event is the breaking of a bond in a molecule. For a single bond this results in the formation of a pair of radicals or a diradical. For a double bond as in diazo compounds or in azides a carbene or a nitrene and nitrogen are formed. All these intermediates will then undergo further mono- or bimolecular dark reactions or eventually recombine to ground state starting materials.

2.1.1 α-Cleavage of Carbonyl Compounds

Photoexcited ketones can undergo the so called Norrish I or α-cleavage reaction wherein one of the bonds to the carbonyl C-atom is broken. Whereas for acyclic ketones a pair of radicals is formed, which then manifests conventional radical chemistry, cyclic ketones give [1, n]diradicals as useful synthetic intermediates [201,202]. Different reactions from these intermediates are observed depending on the number n: for n = 4 (from cyclobutanones), cyclic oxacarbenes (2.1) are formed [203,204], while for n = 5, 6 (from cyclopentanones and cyclohexanones) hydrogen migration leads to either ketenes or unsaturated aldehydes [205]. Obviously these diradicals can close back to starting material; for optically active ketones with the chiral C-atom in α to the CO group this implies racemization (2.2).

(2.1)

The oxacarbene in (2.1) is usually trapped by water or an alcohol to give hydroxy- or alkoxy tetrahydrofuran derivatives. This sequence has been connected with the synthesis of prostaglandines (2.3) [206].

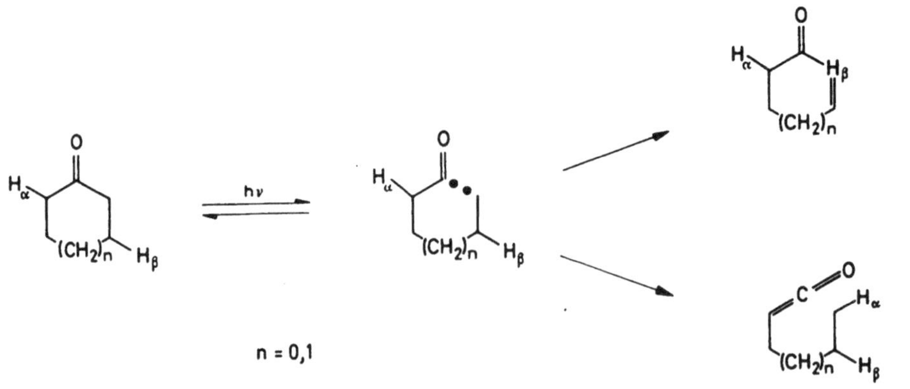

(2.2)

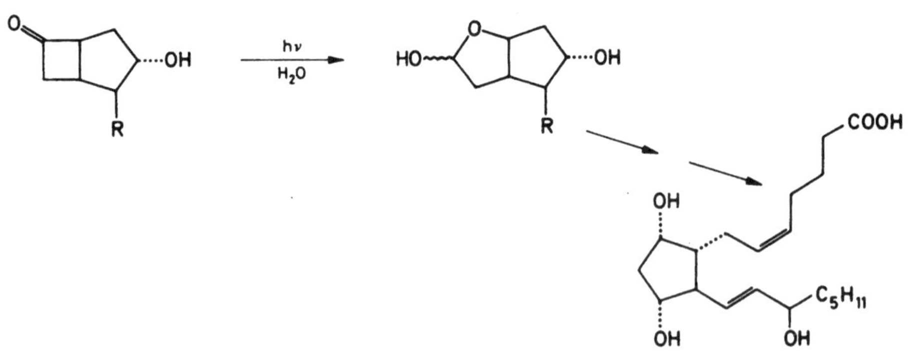

(2.3)

The ketene in (2.2) can be trapped by water or alcohol to give carboxylic acids or esters. Such reactions have already been applied more than seventy years ago in the formation of substituted hexanoic acids (2.4) [207].

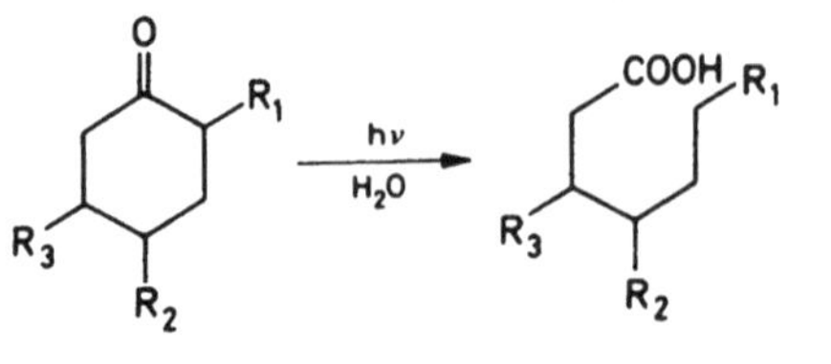

(2.4)

The unsaturated aldehyde in (2.2) very often undergoes intramolecular photocycloaddition with formation of an oxetane (cf. chapter 4.3.6). The overall reaction therefore consists in the formation of a (n + 1)-cyclic oxetane from a n-cyclic ketone, as illustrated in (2.5) [208,209].

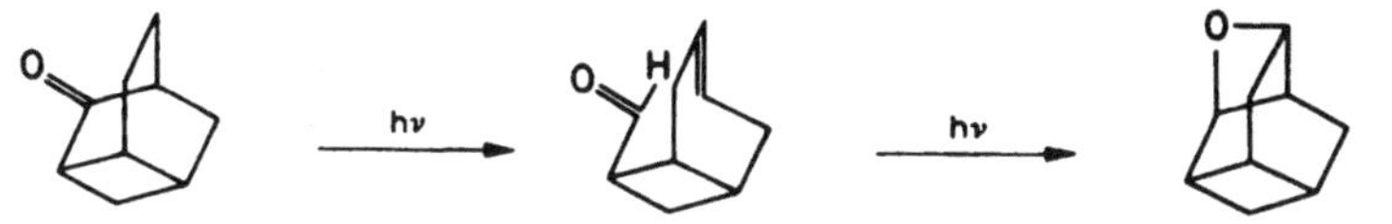

(2.5)

Such a secondary photoreaction can be avoided by trapping the unsaturated aldehyde, e.g. as a hemiacetal. This principle is demonstrated in the synthesis of a furanoid sesquiterpene (2.6) [210].

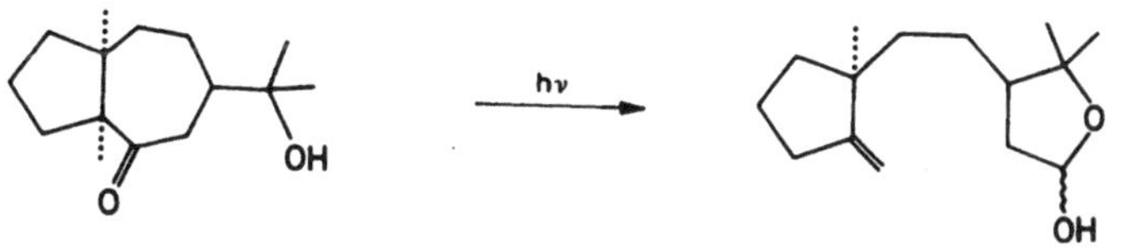

(2.6)

In the presence of proper substituents the primarily formed diradical rearranges to a more stable one, as by a ring opening reaction. In (2.7) medium ring lactones are synthesized by such a sequence [211].

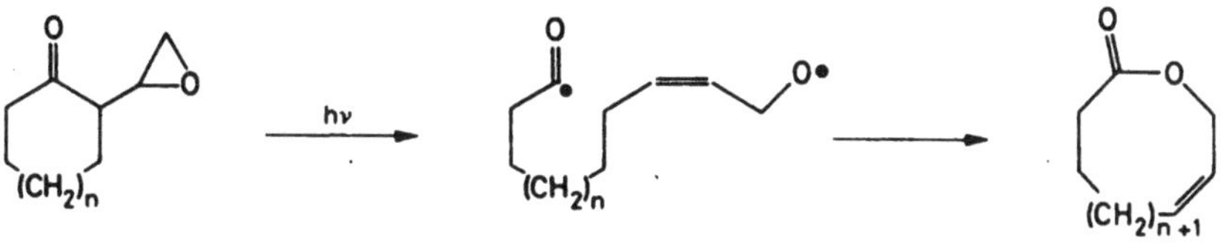

n = 0,1,2

(2.7)

α-Cleavage is also observed in 2-cyclopentenones; the corresponding diradical cyclizes to a cyclopropane derivative (2.8) [212]. The analogous aza-compounds give cyclopropyl carbamates in a similar sequence (2.9) [213].

(2.8)

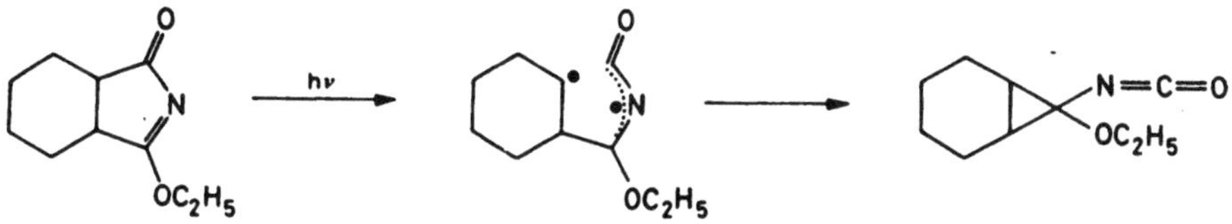

(2.9)

Direct irradiation of β,γ-unsaturated ketones affords rearranged β,γ-unsaturated ketones via a 1,3 shift of the acyl group [214a, b]. This is exemplified in the preparation of a bicyclo[3,3,1]nonanone (2.10) [215].

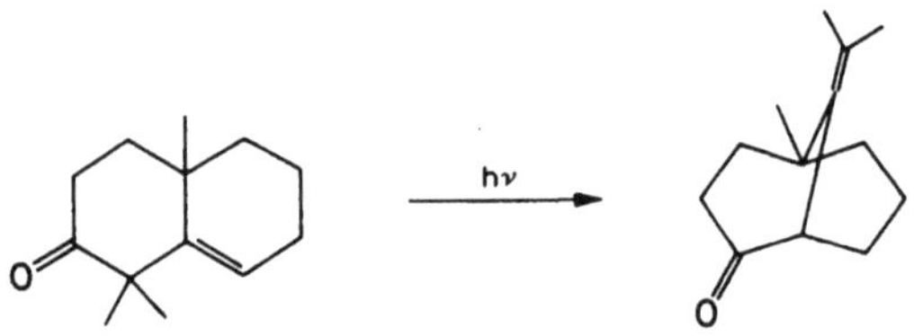

(2.10)

In contrast to 2-cyclohexenones where no α-cleavage is observed this reaction occurs efficiently for 2,4-cyclohexadienones [216, 217]. Among different synthetic applications the synthesis of dimethylcrocetine (a heptaene dicarboxylic acid) may serve as an illustration (2.11) [218].

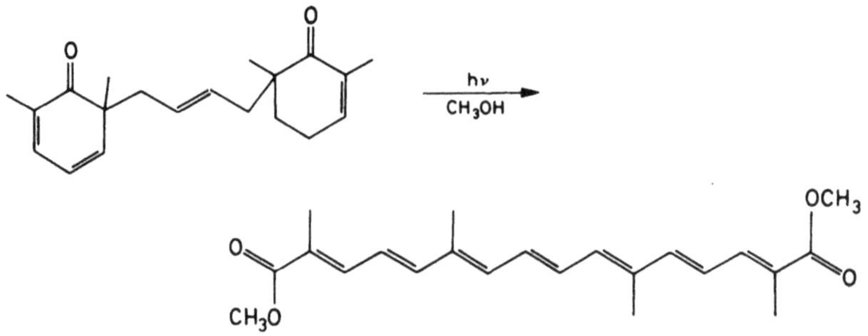

(2.11)

2.1.2 α-Cleavage of Amides and Enamides

Although from the synthetic viewpoint the more important reaction of enamides is the photocyclization, cleavage of the carbonyl C-nitrogen bond can also occur [219]. The reaction product is an enaminoketone as shown in (2.12) [220].

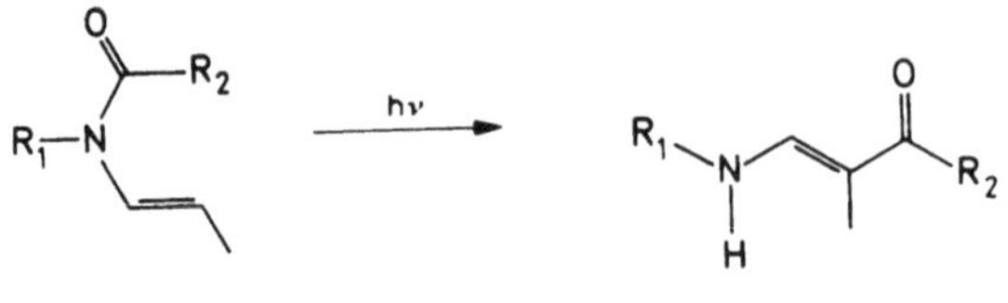

(2.12)

2.1.3 Cleavage of Carbon Halogen Bonds

A bond between carbon and halogen can be cleaved quite easily with light [221]. One of the few synthetic applications of this reaction is the photo-arylation of aromatic compounds, a sequence employed in the synthesis of alkaloids [222, 223] as shown for the key step in the synthesis of boldine (2.13) [223].

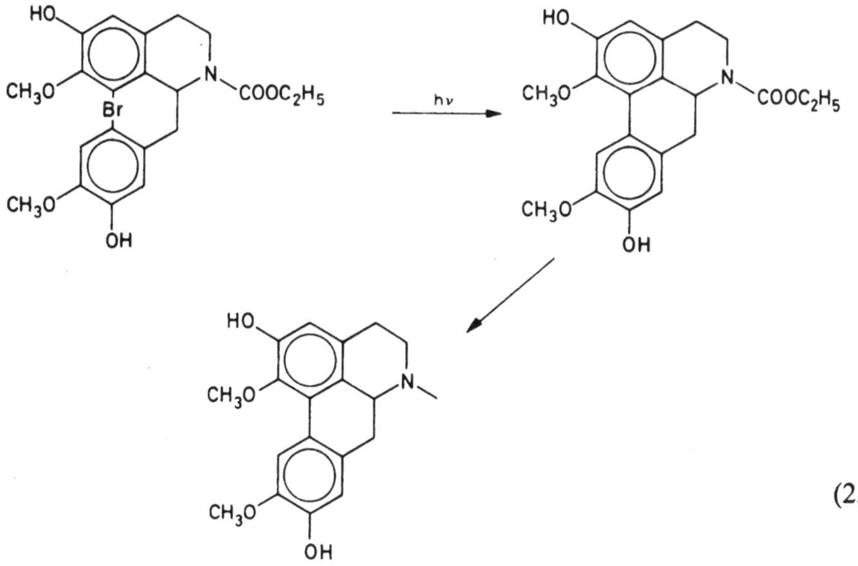

(2.13)

2.1.4 Cleavage of Organic Nitrites

The photochemical cleavage of organic nitrites [224], also known as Barton reaction, has been widely utilized in the steroid field. The mechanistic sequence as described in (2.14) consists of:

a) rupture of the C—*O*—*N*—O bond;
b) abstraction of hydrogen by the oxy-radical;
c) recombination of $NO^{\cdot}$ and the alkyl radical with formation of a C-nitroso compound which tautomerizes to an oxime.

The reaction has been used in the synthesis of cortisone oxime [225] (2.15) and β-amyrene [226].

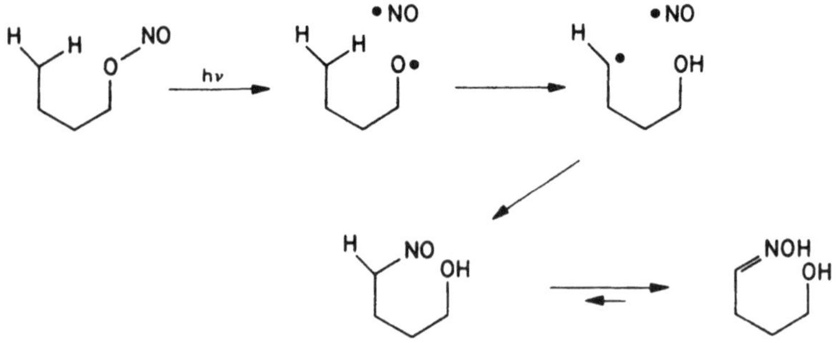

(2.14)

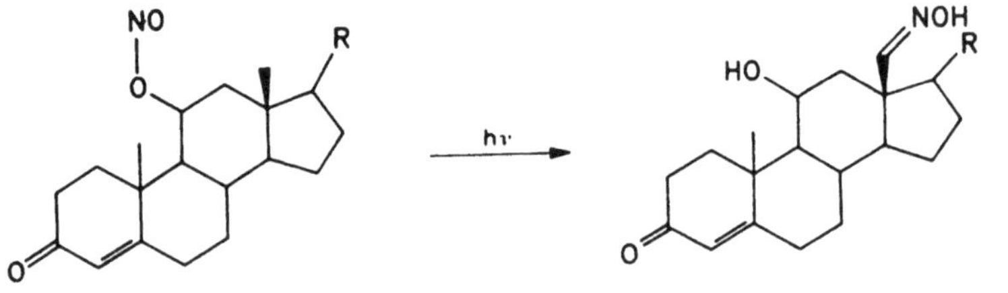

2.15)

2.1.5 Cleavage of Diazo Compounds

Diazoalkanes and diazocarbonyl compounds are decomposed by light to give molecular nitrogen and a carbene [227, 228]; the ketocarbenes very often rearrange to ketenes. The most important application of this photochemical generation of carbenes is the synthesis of cyclopropanes. Singlet carbenes usually add to olefins with retention of configuration of the olefin, while triplet carbenes add in a distinct two-step sequence leading to Z- and E-cyclopropanes [229]. There are exceptions to this rule as e.g. singlet fluorenylidene [230] which adds to Z- or E-butene in a nonstereospecific manner. Typical reactions are outlined below: in (2.16) a pentacyclic spiro-fused system is obtained [231]; in (2.17) the cyclopropane is formed in 100% yield, but more highly substituted olefins tend to give products arising from carbene insertion into an allylic C—H bond [232]. Use of this sequence has also been made in the synthesis of menthofuran, a furanoid terpene (2.18) [233].

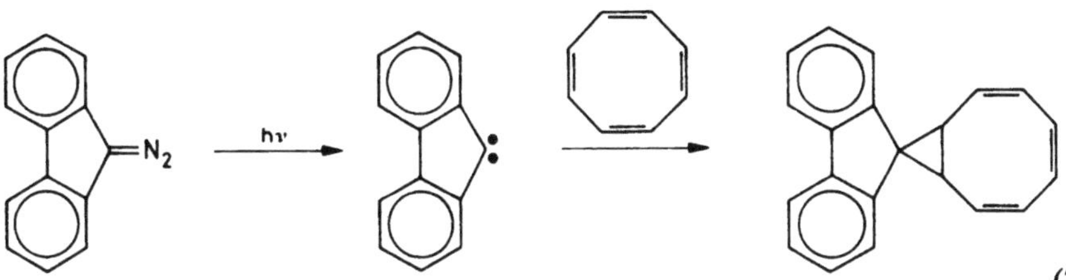

(2.16)

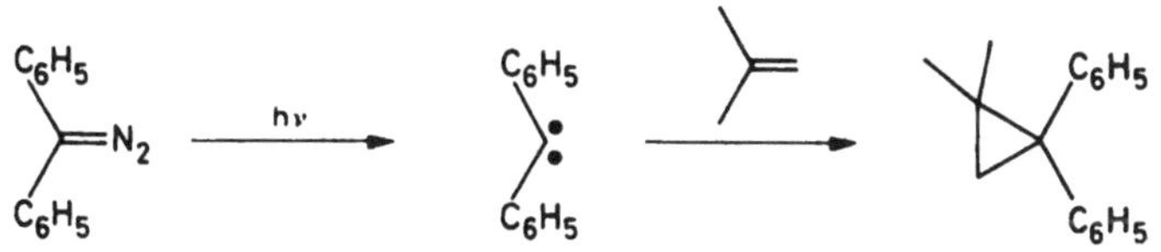

(2.17)

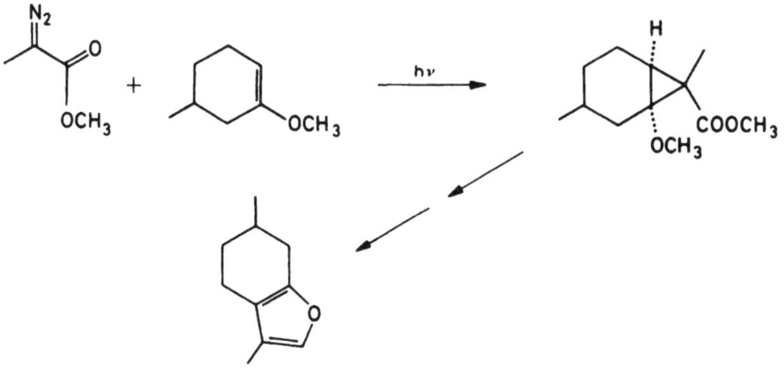

(2.18)

Under special conditions carbenes have been observed to dimerize to alkenes as illustrated in the synthesis of tetramesitylethylene (2.19) [234].

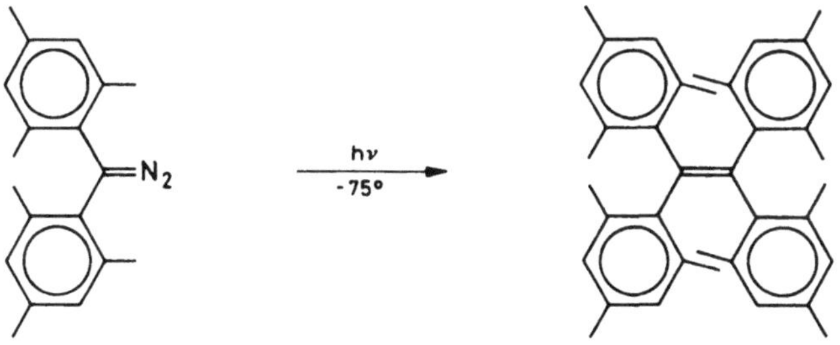

(2.19)

Ketocarbenes generated from diazoketones give two main type of reactions. The first one is the "conventional" carbene reaction, i.e. intramolecular insertion into a C—H or C—C bond, as applied in the synthesis of a tricyclopentanone [235] (2.20), in the key step of the synthesis of the alkaloid (±)glaziovine [236] (2,21) or in the synthesis of a β-lactam (2.22) [237].

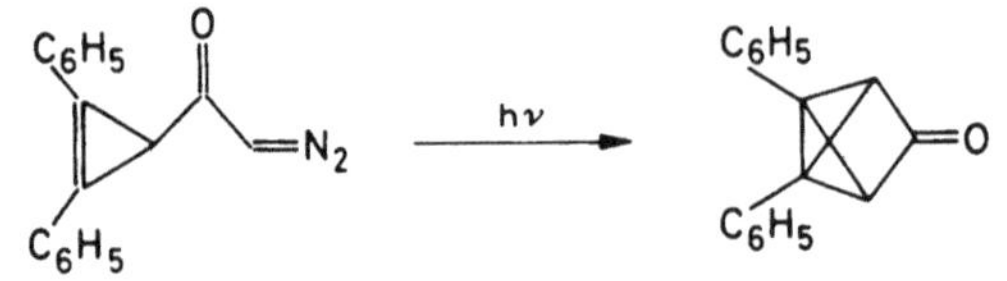

(2.20)

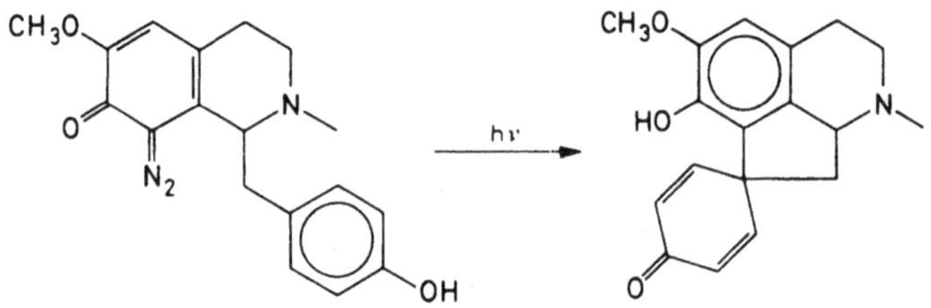

(2.21)

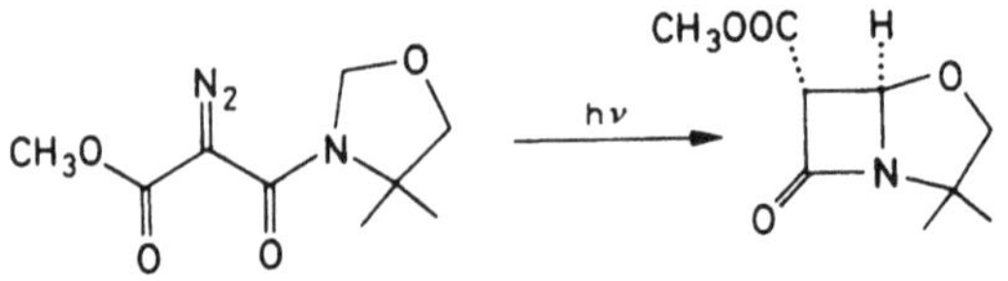

(2.22)

The second reaction mode is rearrangement of the ketocarbene to a ketene. In the presence of a C—C double bond this species reacts further via an intramolecular photocycloaddition (cf. chapter 4.3.3), as shown in (2.23) [238].

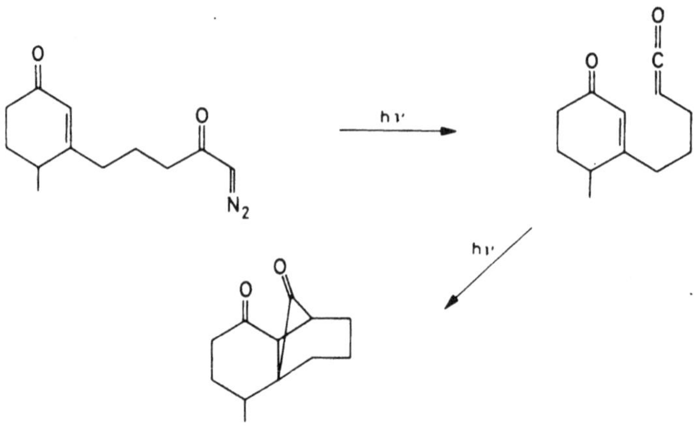

(2.23)

The photolysis of cyclic diazo ketones in hydroxylic solvents leads to ring contracted carboxylic acid derivatives *via* this ketocarbene → ketene rearrangement. Examples of such reactions are given in (2.24) [239] and (2.25) [240]. In this last example a photoequilibrium between the diazo ketone and its valence isomer, a diazirine, has been observed, both products then eliminating nitrogen to afford the cyclobutane carboxylic acid.

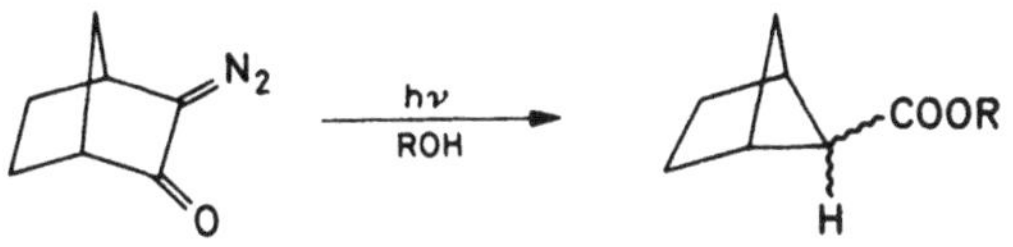

(2.24)

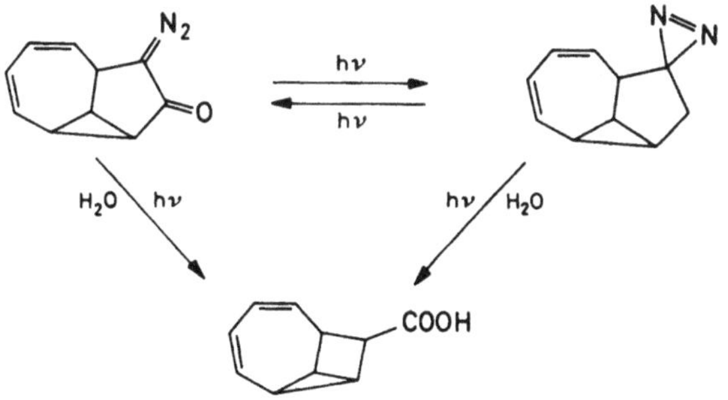

(2.25)

2.1.6 Cleavage of Azides

Organic azido compounds behave like diazo compounds in photochemical reactions. Again molecular nitrogen is eliminated and products deriving from nitrenes are formed. While the intermediacy of aryl- or acyl nitrenes in such reactions has been unambiguously proven, there is no evidence for the occurence of alkyl nitrenes; instead, in the photolysis of alkyl azides a mechanism wherein nitrogen elimination and alkyl group migration occur synchronously has been suggested [241a, b, 242].

The photolysis of tertiary alkyl azides has been utilized for the synthesis of mono- and di-N-derivatives of biacetyl (2.26) [243a, b]

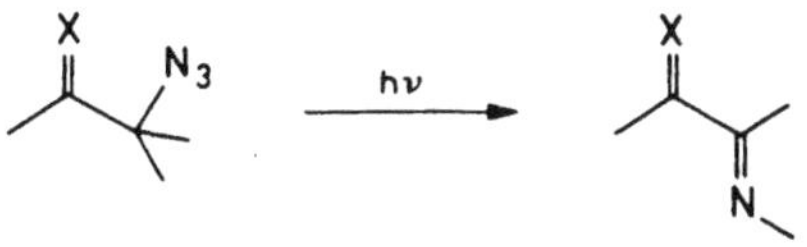

(2.26)

X = O, NOH, NCH3

Use of the insertion of an arylnitrene into a C—H bond has been made of in the synthesis of 9H-pyrimido[4,5-*b*]indole (2.27) [244].

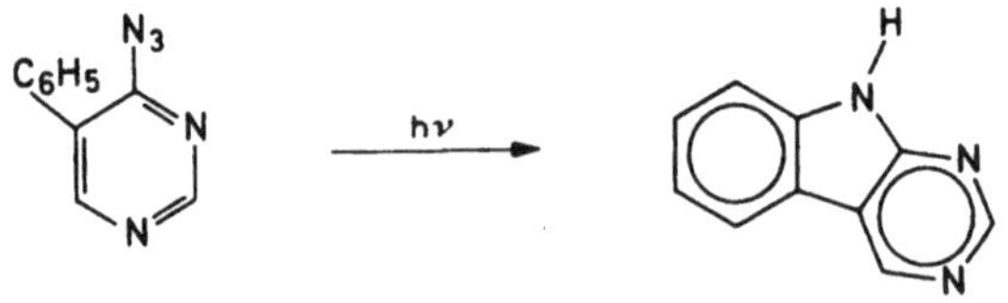

(2.27)

The photolysis of acyl azides affords lactames (2.28) [245]

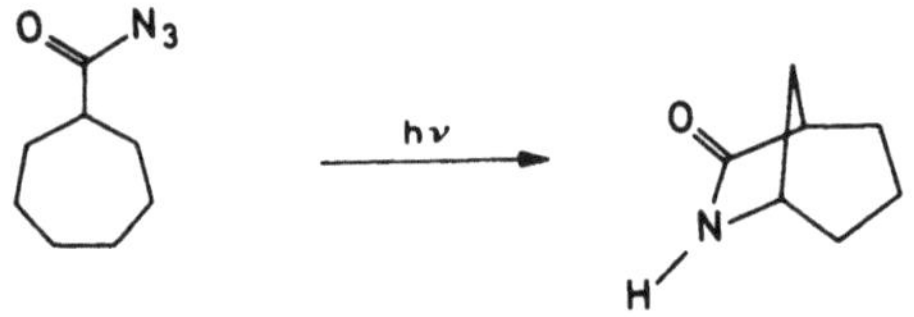

(2.28)

Azidoformates have been used as starting material for the synthesis of 1,3-oxazine derivatives (2.29) [246].

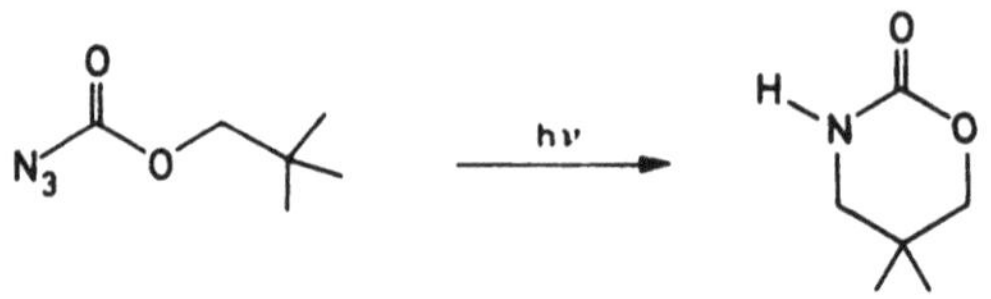

(2.29)

Vinyl azides decompose to 3H-azirines. This reaction has been reviewed recently [247].

2.2 Photoextrusion Reactions

The initial step in a photochemical extrusion reaction is again cleavage of a bond to give a pair of radicals or a diradical, but now a second bond is broken immediately after the first one, thus leading to the extrusion of a small molecule as CO, CO_2, N_2 or SO_2. As in a recent review on photoextrusion of small molecules [248] only such reactions will be outlined wherein the small molecule that is to be expelled is initially part of a ring, and where the product formed is again a cyclic system.

2.2.1 CO-extrusion from Carbonyl Compounds

As already discussed, cyclobutanones afford oxacarbenes on direct irradiation (cf. chapter 2.1.1). In sensitized irradiations these compounds decarbonylate to afford cyclopropanes (2.30) [249].

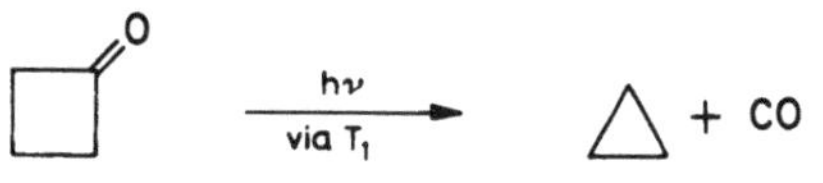

(2.30)

Synthetic applications of other decarbonylation reactions are found in the conversion of cyclooctatetraene to barrelene [250], with the photodecarbonylation of a Diels-Alder adduct as key step (2.31) and the preparation of tetrathioesters from 1,3-dithioles (2.32) [251]. The most remarcable application of such a reaction up to date is the synthesis of tetra t.butyltetrahedrane from a tricyclic ketone precursor (2.33) [252].

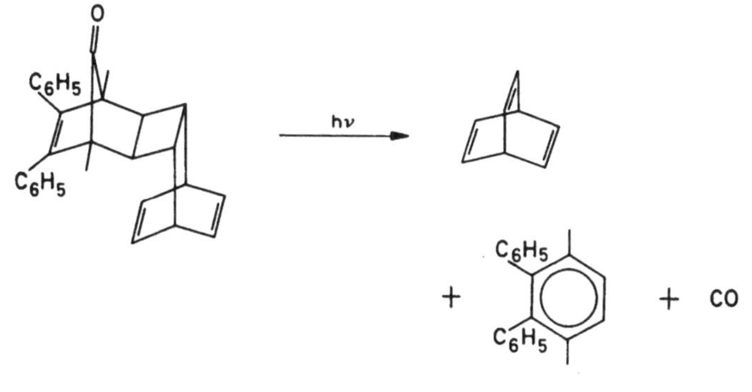

(2.31)

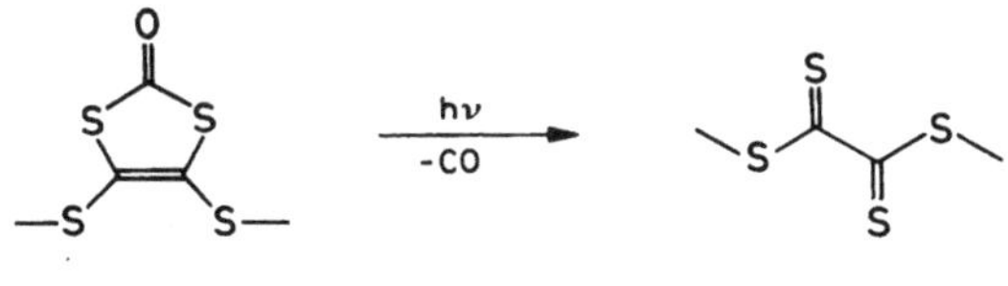

(2.32)

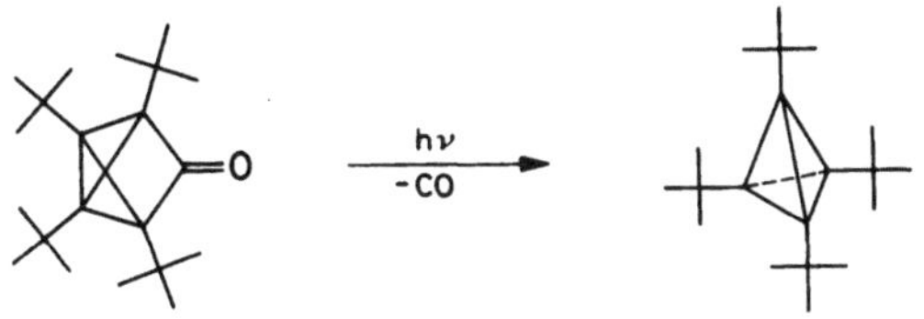

(2.33)

2.2.2 CO_2-extrusion from Carboxylic Acid Derivatives

The photolysis of carboxylic acids and derivatives as lactones, esters and anhydrides can yield decarboxylated products [253]. This reaction has been utilized in the synthesis of α-lactones from cyclic diacyl peroxides [254] (2.34) and in the synthesis of [2,2]paracyclophane by bis-decarboxylation of a lactone precursor (2.35) [255]. This latter product was also obtained by photoinduced desulfurization of the analogous cyclic sulfide in the presence of triethyl phosphite [256].

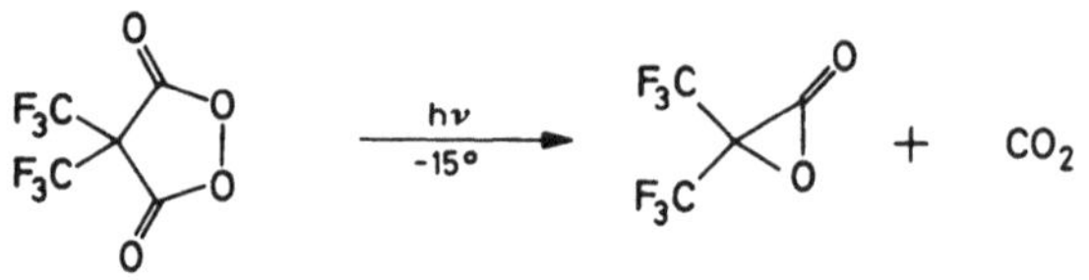

(2.34)

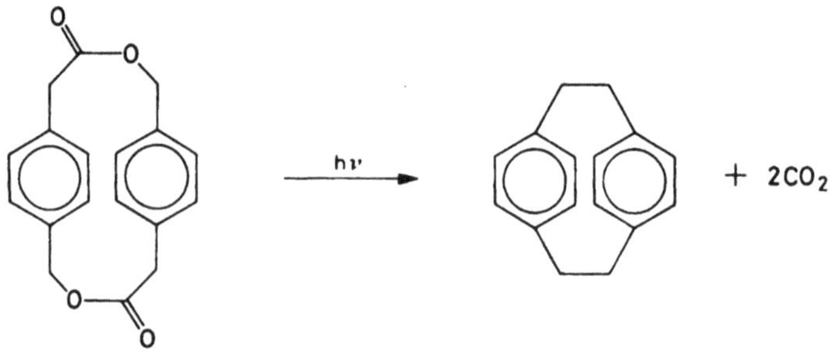

(2.35)

2.2.3 N_2-extrusion from Cyclic Azo Compounds

The irradiation of aliphatic azo compounds usually leads to the formation of molecular nitrogen and a pair of radicals, or for cyclic compounds, a diradical [257]. Many cyclic azoalkanes are decomposed efficiently by irradiating in the low-energy n—π* absorption band, with the exception of compounds containing the azo linkage in a six membered ring. These so called "reluctant" azoalkanes undergo facile denitrogenation with light of $\lambda = 185$ nm [258a, b].

Photolysis of an azo compound provided the first synthesis of prismane, albeit in very low yield (2.36) [259].

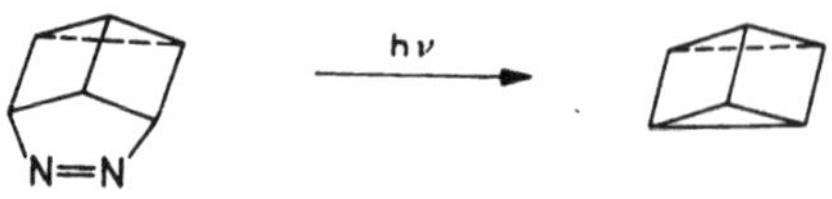

(2.36)

Other applications are found in the synthesis of triquinacene (2.37) [260] and of benzocyclopropenes (2.38) [261].

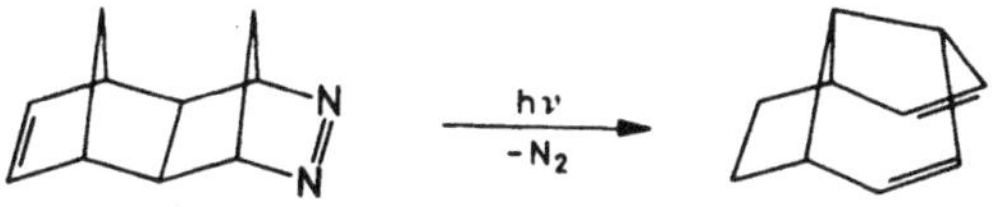

(2.37)

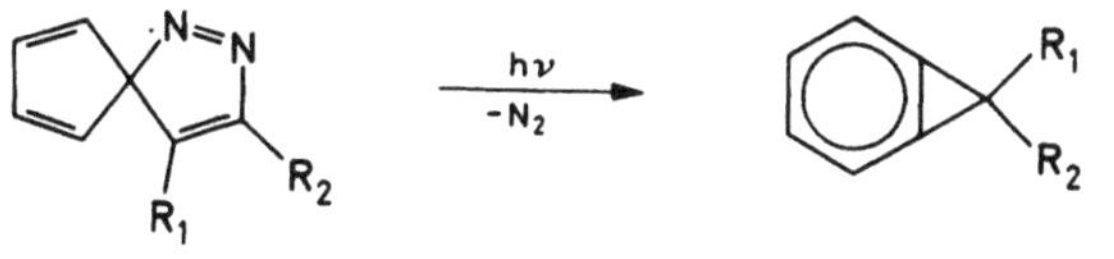

(2.38)

Other examples of strained polycyclic hydrocarbons formed by photolysis of an azo precursor are spiro[2,2]pentane [262] (2.39), housane [263] (2.40) and exotricyclo[3.2.1.0^{2,4}]octene (2.41) [264].

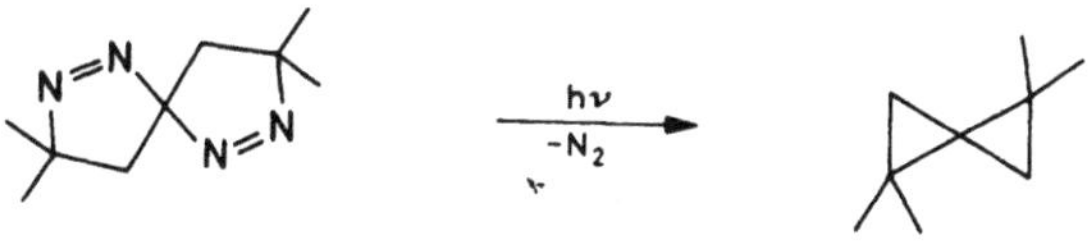

(2.39)

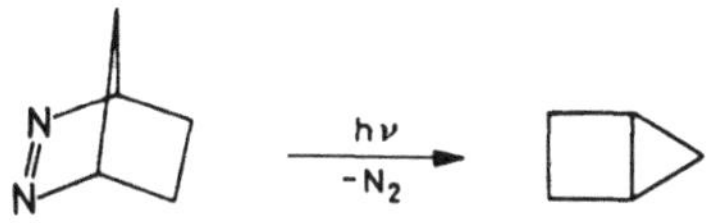

(2.40)

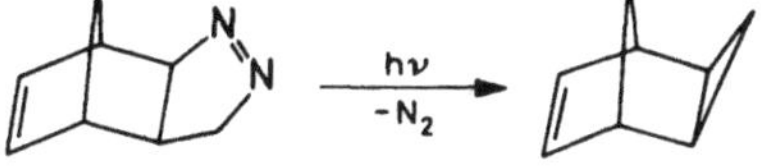

(2.41)

3 Photochemical Rearrangement Reactions in Synthetic Organic Chemistry

3.1 Photorearrangements

This chapter has to do with different types of photoisomerization reactions. This includes the formation of constitutional, configurational, conformational and valence isomers.

3.1.1 E—Z Isomerization of Alkenes

One important aspect of the photochemistry of alkenes [301,302] is the E—Z isomerization around the C—C double bond [303]. This is also valid for cycloalkenes with the obvious exception of cyclopropenes [304a, b] which exhibit a distinct photochemical behaviour, and cyclobutenes and cyclopentenes where the ring is to rigid to allow sufficient twisting of the double bond.

The photochemical isomerization of E-stilbenes has been applied in the preparation of phenanthrenes, as Z-stilbenes undergo electrocyclic ring closure (cf. chapter 3.1.3) to dihydrophenanthrenes which in turn are easily oxidized to phenanthrenes (3.1) [305]. This sequence has also been employed in the synthesis of benzoquinolines [306] or of benzoquinolizines (3.2) [307].

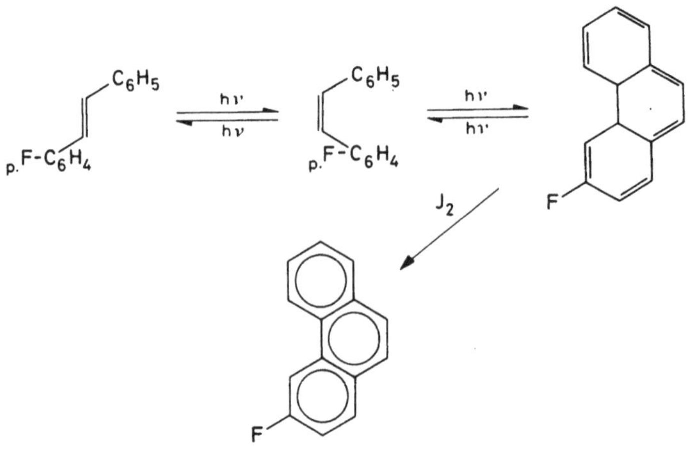

(3.1)

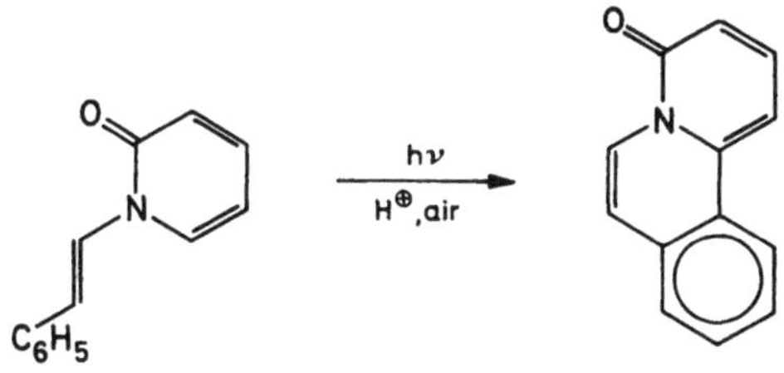

(3.2)

Generally for simple alkenes the E-isomer is thermodynamically more stable than the Z-isomer. Photochemical isomerization is thus a powerful method to obtain Z-alkenes in good yields, as shown for β-ionol (3.3) [308].

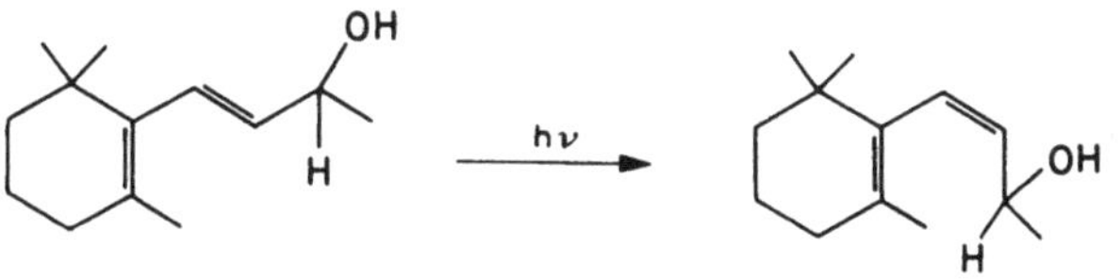

(3.3)

Another important feature of this reaction is the synthesis of highly strained E-cycloalkenes. Thus E,Z-cycloocta-1,3-diene has been obtained from the Z,Z-compound in sensitized irradiation [309] (3.4), while E-cyclooctene was obtained from the Z-isomer by direct irradiation (3.5) [310]. The synthesis of a trans doubly bridged olefin has also been reported (3.6) [311].

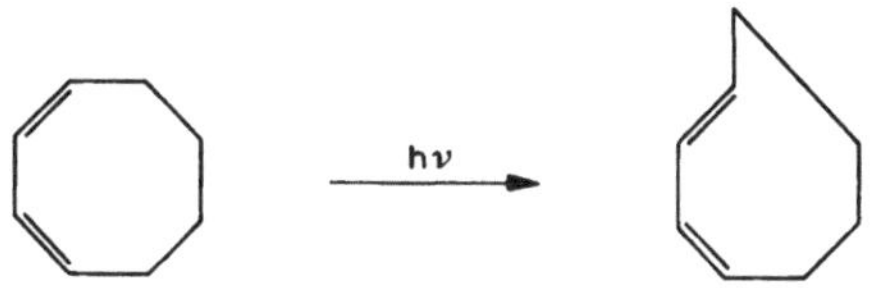

(3.4)

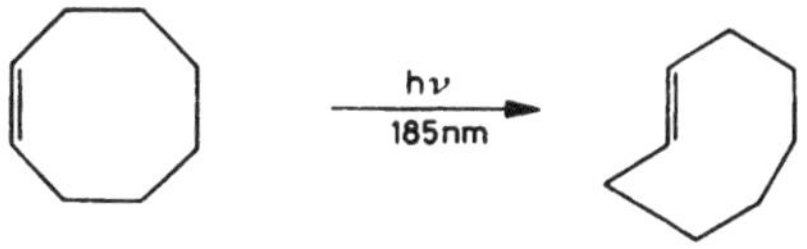

(3.5)

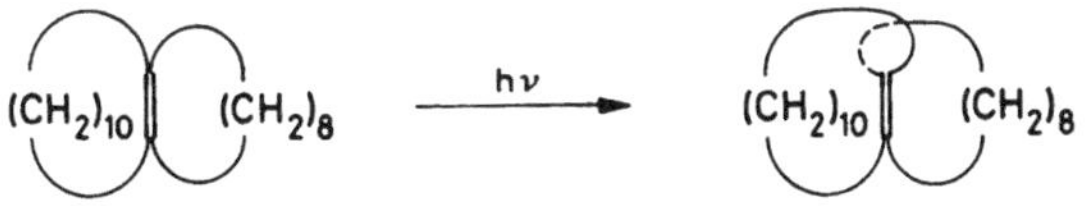

(3.6)

The E-isomers of cycloheptenes and cyclohexenes are usually not stable enough to be isolated; nevertheless they can be trapped by alcohols, as for cycloheptenone [312] (3.7), or observed spectroscopically as for phenylcyclohexene. This latter olefin also undergoes photoaddition to alcohols, but via electron transfer from the alkene to the excited sensitizer, the resulting radical cation then reacting with the solvent (3.8) [313].

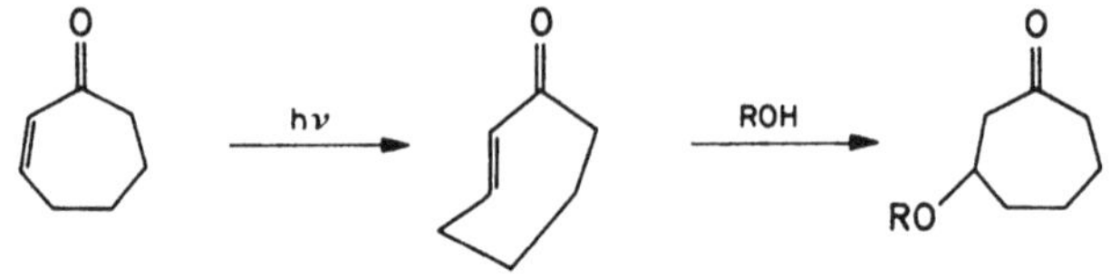

(3.7)

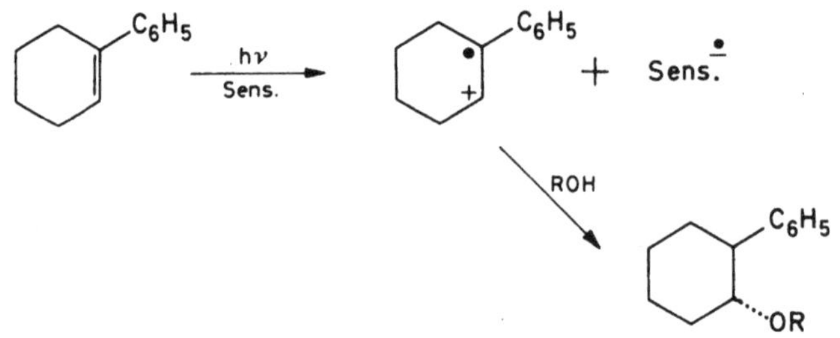

(3.8)

3.1.2 Electrocyclic Reactions of Conjugated Dienes and Trienes

Electrocyclic reactions were first described by Woodward and Hoffmann in their classic series of articles. One very interesting aspect of such reactions is, that for a given conjugated polyene photochemical transformation leads to the opposite stereochemical outcome than the thermal one [314].

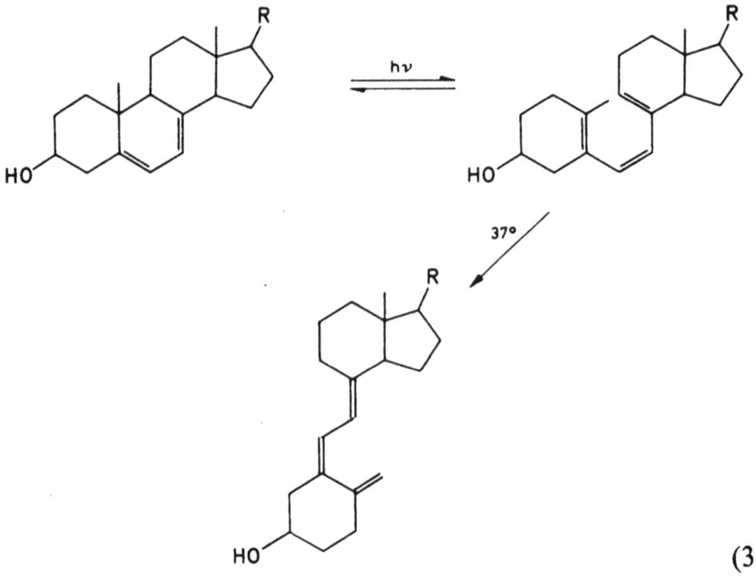

(3.9)

The photoequilibrium between 1,3-cyclohexadienes and 1,3,5-hexatrienes [315,316] is the key step in the synthesis of vitamin D, as shown in the formation of vitamin D_3 ($R = C_8H_{17}$) via a [1,7]sigmatropic H-shift from previtamin D which is obtained by irradiating provitamin D (3.9) [317].

The ring closure of 1,3-butadienes to cyclobutenes has been utilized for synthetic purposes as in the preparation of cyclobutene itself [318] (3.10), of a bicycloheptenone [319] (3.11), of a bicyclic lactam [320] (3.12) or of diphenylbicycloheptene (3.13) [321].

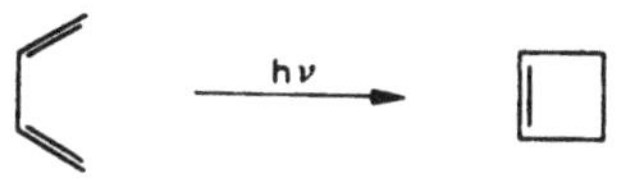

(3.10)

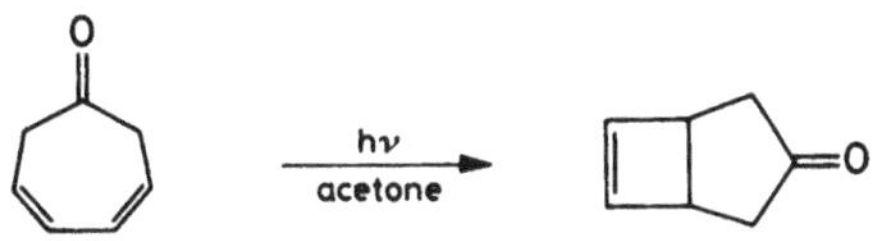

(3.11)

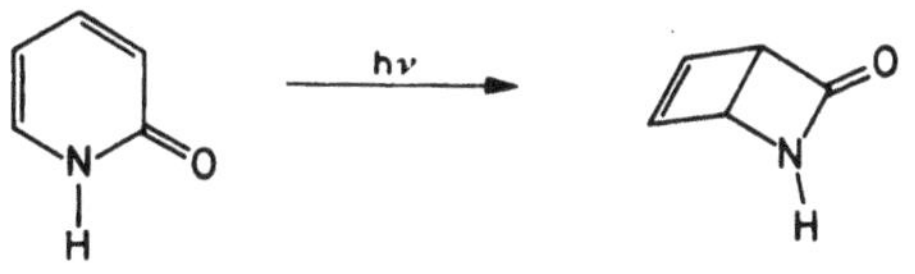

(3.12)

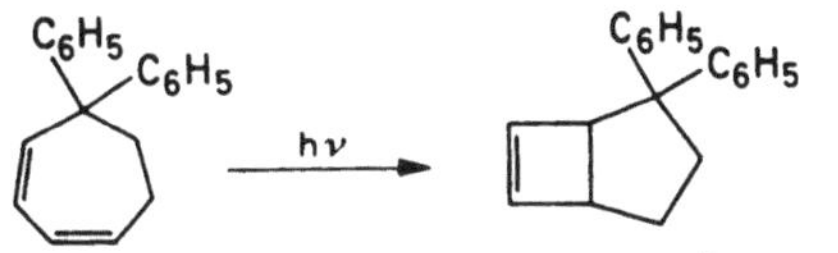

(3.13)

3.1.3 Di-pi-methane Rearrangement of 1,4-Dienes

1,4-Dienes can be converted photochemically to vinyl cyclopropanes. This reaction, also named the Zimmerman rearrangement [322], is common to molecules in which two π moieties are substituents on a "methane" carbon [323]. The skeletal changes follow the mechanistic sequence in (3.14).

(3.14)

Early applications of this reaction are found in the conversion of barrelene to semibullvalene [324] (3.15) or in the preparation of an azabullvalene (3.16) [325]. In a similar reaction azabarbaralene has been prepared from aza-bicyclononatriene [326].

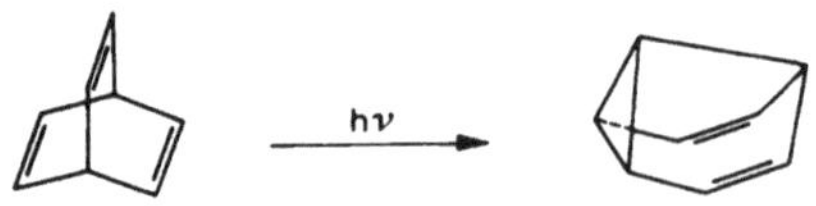

(3.15)

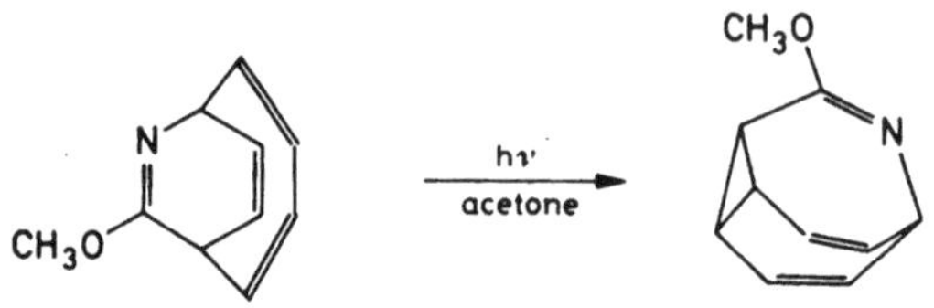

(3.16)

The di-π-methane rearrangement is also a convenient way of obtaining polycyclic fused ring systems as illustrated in the synthesis of a tricyclo-undecane (3.17) [327]. In the irradiation of dihydrotriquinacene the initial bonding scheme is identical as in (3.14) but ultimate cyclopropane formation is hindered by structural reasons (3.18) [328].

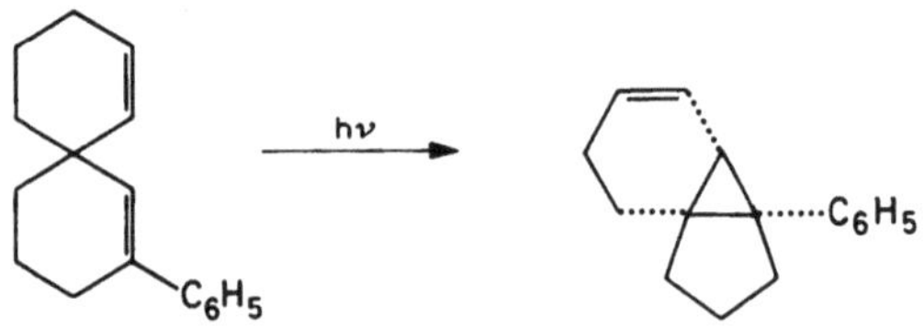

(3.17)

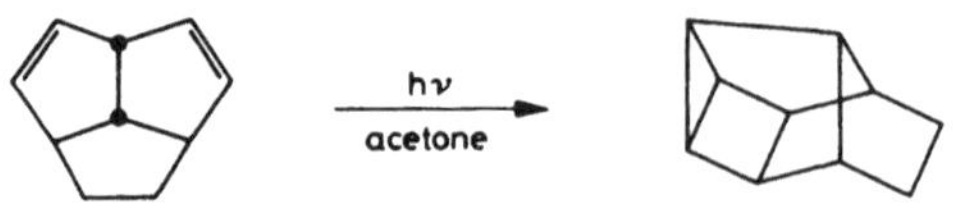

(3.18)

One of the C—C double bonds can also be part of an enone system as shown in the synthesis of 2-cyclopropyl-2-cyclopentenones. Such a photo-product was then converted to the sesquiterpene taylorione (3.19) [329].

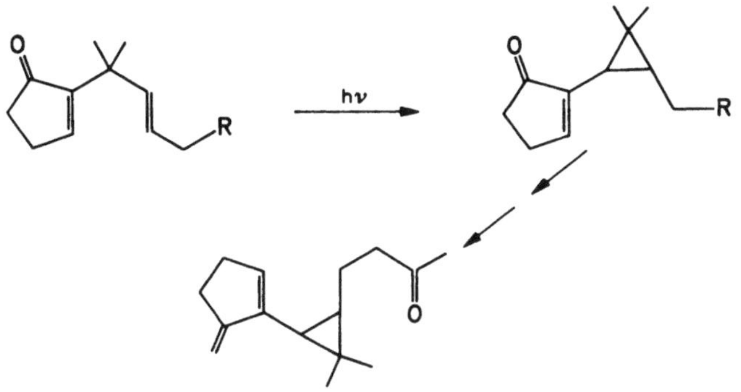

(3.19)

3.1.4 Photorearrangement of Arenes

Benzene is commonly used as solvent in photochemical reactions; nevertheless one has to take into consideration that arenes do undergo photorearrangements [330a, b]. Benzene isomers as Dewar benzene [331] (3.20) or benzvalene [332] (3.21) have been prepared by photolysis of benzenes.

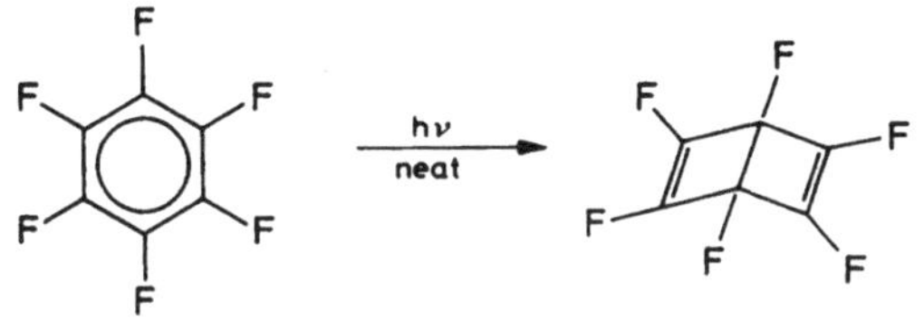

(3.20)

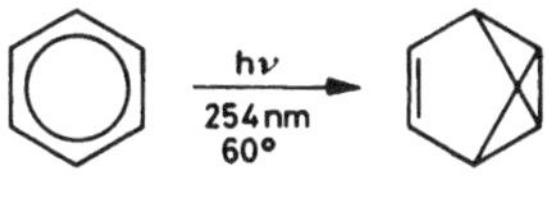

(3.21)

3.1.5 Photorearrangement of α,β- and β,γ-Unsaturated Ketones

Enones undergo a variety of photorearrangements [333]. In contrast to α,β-unsaturated ketones which react in many different ways, β,γ-unsaturated ketones undergo the oxadi-π-methane rearrangement in sensitized irradiations; in complete analogy the products formed are cyclopropyl ketones [334–336].

Irradiation of 4,4-dialkyl-2-cyclohexenones affords bicyclo[3.1.0]hexanones in a so called "lumiketone" rearrangement (3.22) [337]. The reaction proceeds with inversion of configuration on the disubstituted C-atom (3.23) [338].

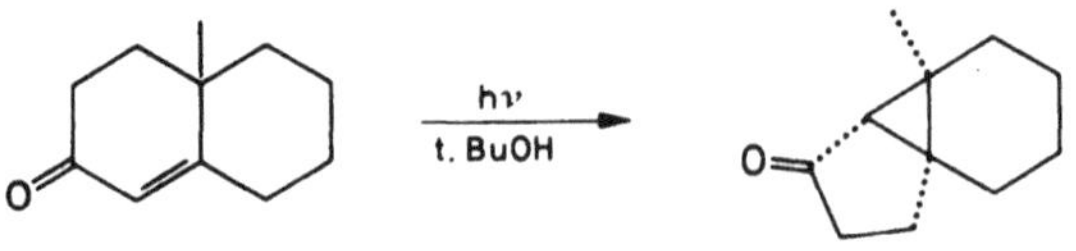

(3.22)

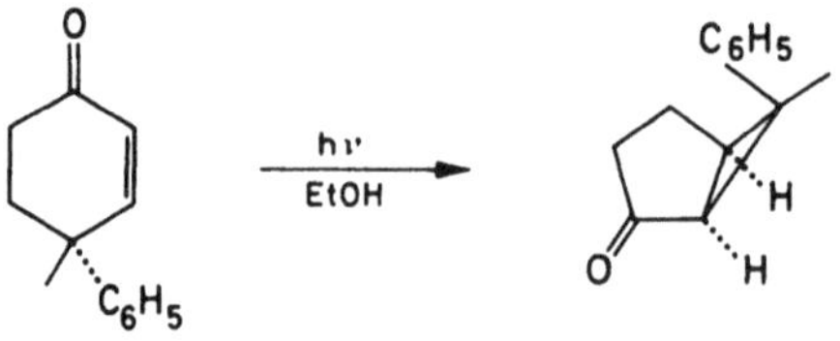

R-(+) (-)

(3.23)

2-cyclopentenones may isomerize to 3-cyclopentenones as shown for the propellane molecule in (3.24). Cleavage of the cyclobutane C—C bond β to the carbonyl group affords a diradical with an allylic centre, the recombination then giving a rearranged tricycloundecenone [339]. The arrows in (3.24) indicate the cleaved and the newly formed C—C bond.

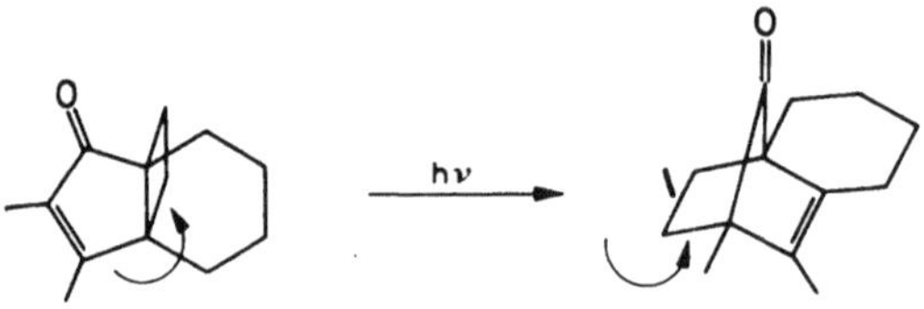

(3.24)

Cyclopentenyl ketones can undergo photoarylation when irradiated in the presence of a Lewis acid catalyst (3.25) [340].

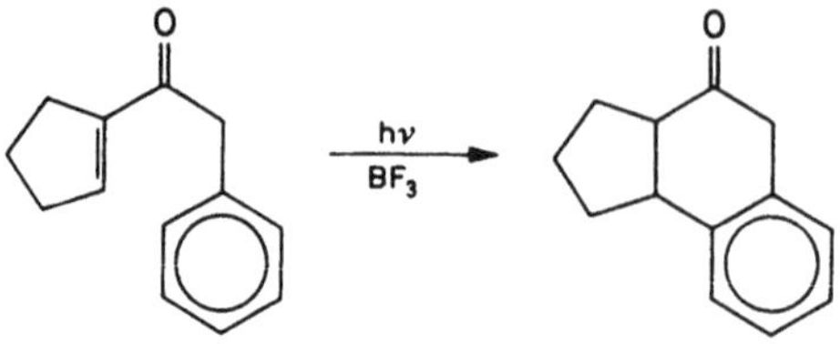

(3.25)

The bicyclooctadienone in (3.26) is converted into a cyclopropyl ketene via a [3,3]sigmatropic rearrangement [341].

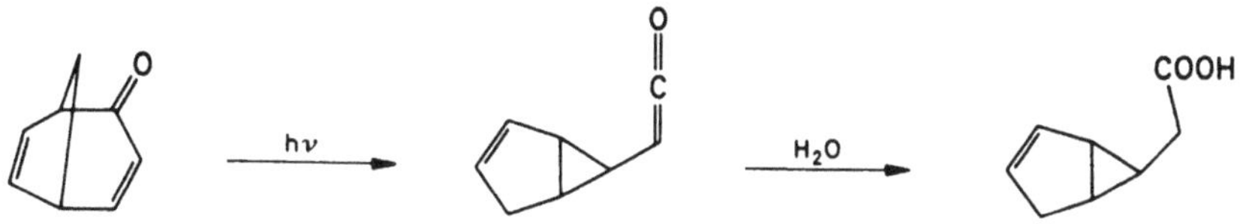

(3.26)

As mentioned earlier (cf. chapter 2.1.1), β,γ-enones undergo an 1,3 acyl shift from the S_1-state. In sensitized reactions they rearrange to cyclopropyl ketones in the so termed oxadi-pi-methane rearrangement, as illustrated for the naphthalenone in (3.27) [342].

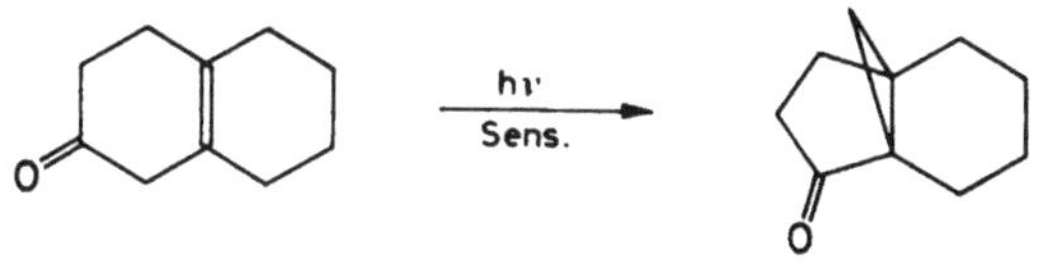

(3.27)

The analogous rearrangement of the bicyclooctenone in (3.28) gives a product which serves as a synthon for polycyclopentanoid terpenes and prostacyclins [343a, b].

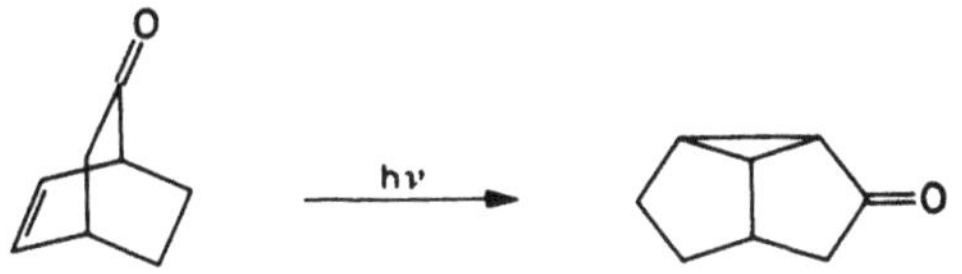

(3.28)

3.1.6 Photorearrangement of 2,5-Cyclohexadienones

The rearrangement of cross-conjugated cyclohexadienones to bicyclo[3.1.0]-hex-3-en-2-ones was one of the first photochemical reactions to be thoroughly studied for mechanistic and synthetic purposes [344]. The examples outlined below are the conversion of the lactone α-santonin to lumisantonin [345] (3.29) and the synthesis of a precursor of the spirocyclic sesquiterpene β-vetivone (3.30) [346].

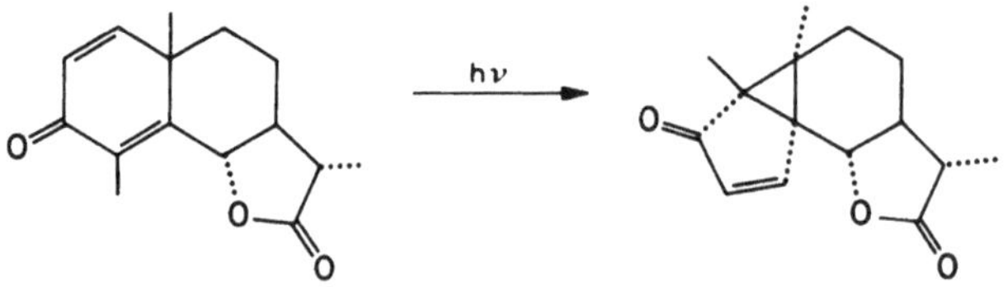

(3.29)

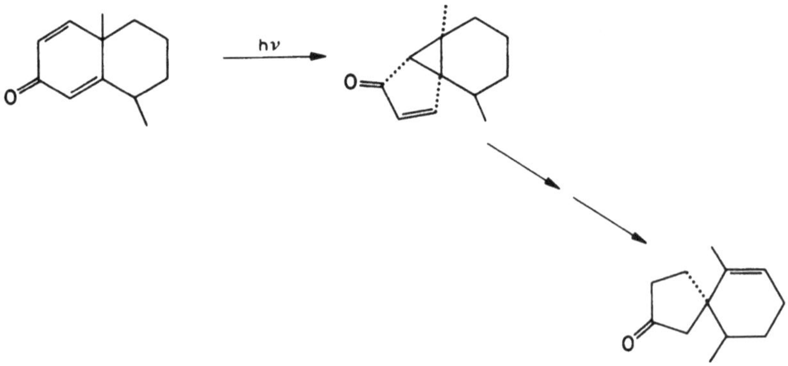

(3.30)

In acetic acid the intermediate zwitterion is trapped by acetate affording cyclopentenones with an acylated hydroxymethylene group on C-4. Use of this sequence has been made of in the synthesis of the sesquiterpenes cyclocolorenone [347] (3.31) and α-cadinol (3.32) [348].

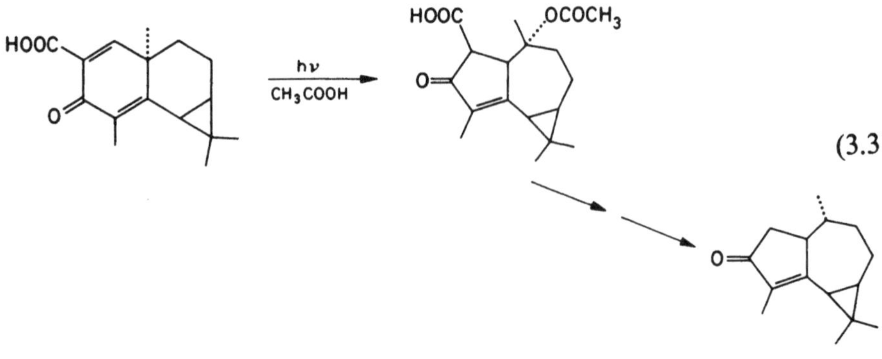

(3.31)

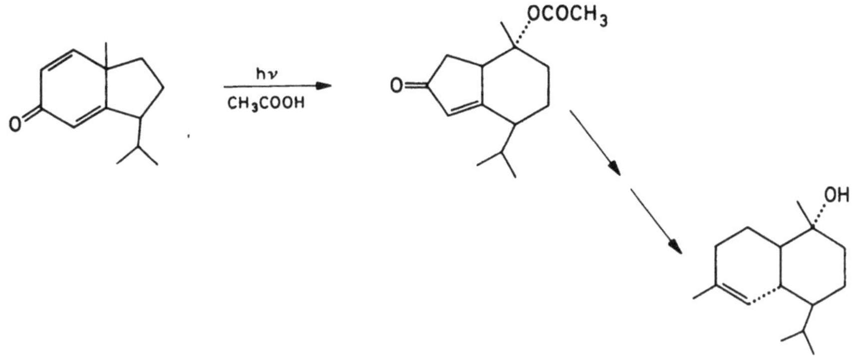

(3.32)

3.1.7 Photorearrangement of Heterocycles

There are obviously no common mechanistic features for the photorearrangement of heterocycles in general. Yet three-membered rings [349)] and five-membered ring heterocycles [350)] exhibit a quite particular behaviour. This topic has also been reviewed in a heterocyclic chemistry series [351)]. In this chapter only some selected reactions of preparative interest will be outlined.

Oxonin has been obtained in the direct or sensitized irradiation of cyclooctatetraeneoxide (3.33) [352)].

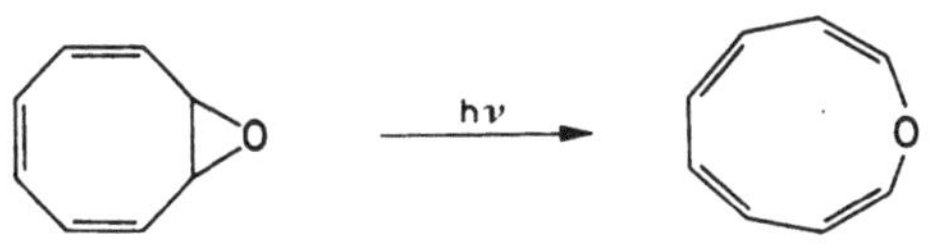

(3.33)

Dihydropyrazines can be converted in good yields to imidazoles (3.34) [353)].

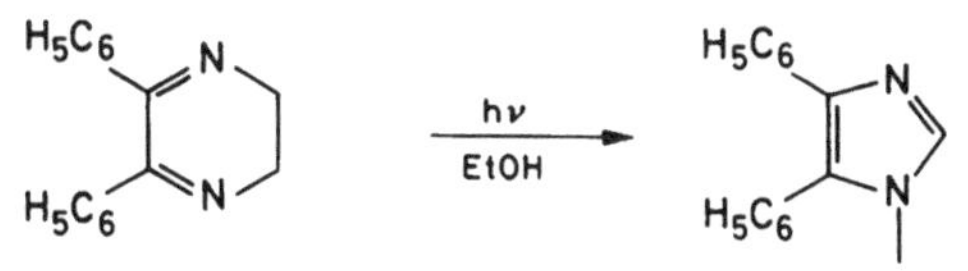

(3.34)

1,2-Diazepines are formed in almost quantitative yield in the irradiation of pyridinium ylides (3.35) [354)].

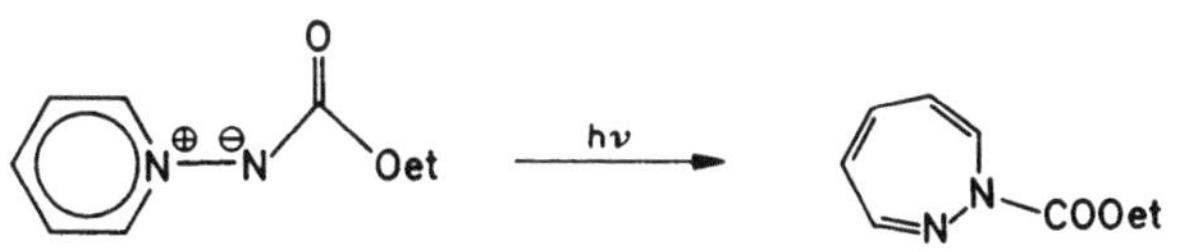

(3.35)

The 1,3-dipolar intermediate obtained in the ring opening reaction of 3H-azirines via cleavage of the C—C bond has been trapped to give oxazolines [355)] (3.36) and pyrrolines (3.37) [356)].

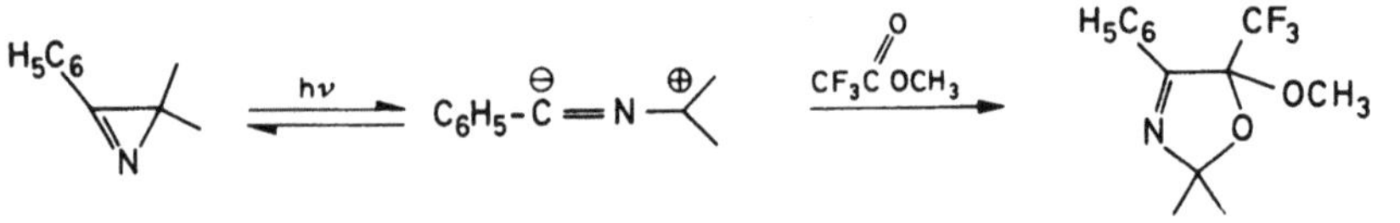

(3.36)

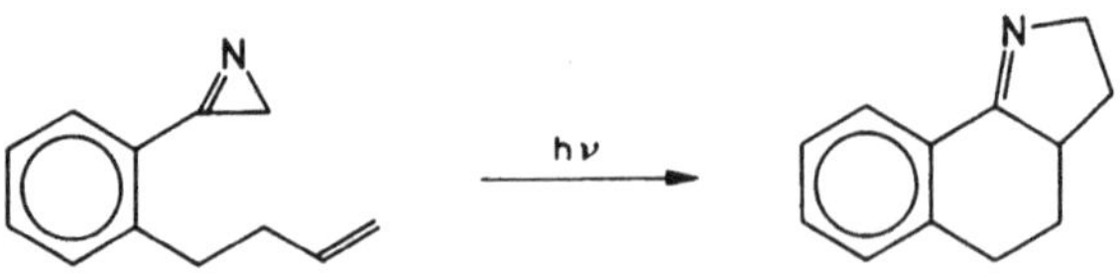

(3.37)

Electrocyclic ring opening of the aziridine in (3.38) affords aza[17]annulene [357].

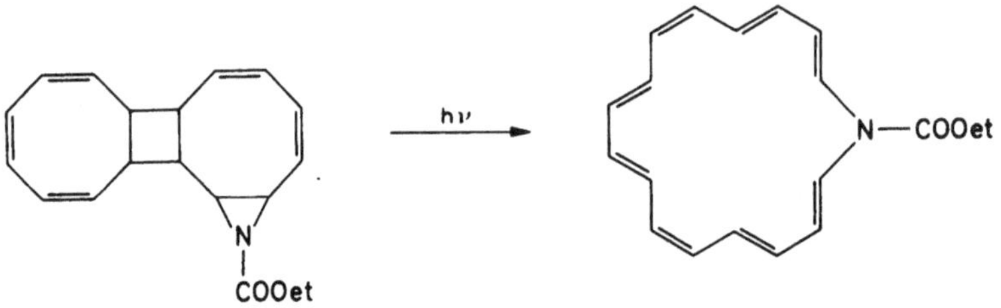

(3.38)

3.1.8 Photocyclization of Enamides

The photocyclization of enamides to quinolines or isoquinolines has become an important reaction in the synthesis of alkaloids [219,358]. It has recently been applied in the preparation of the isoquinoline alkaloid polycarpine [359]. The principle of the reaction is demonstrated in the preparation of dihydroquinolines [360] (3.39) and of spirocyclohexaneisoquinoline derivatives (3.40) [361]. In each case the electrocyclic ring closure product undergoes a subsequent 1,5-hydrogen shift.

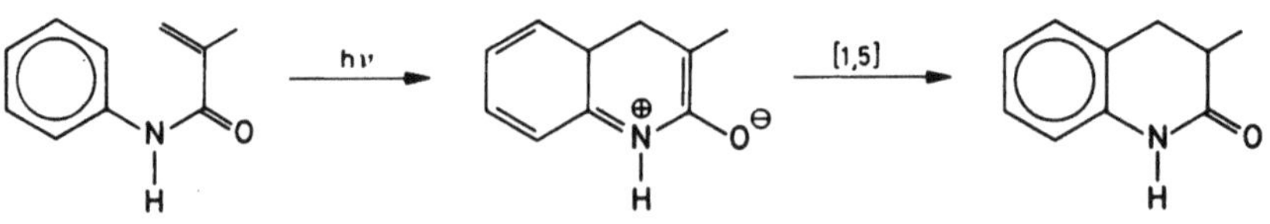

(3.39)

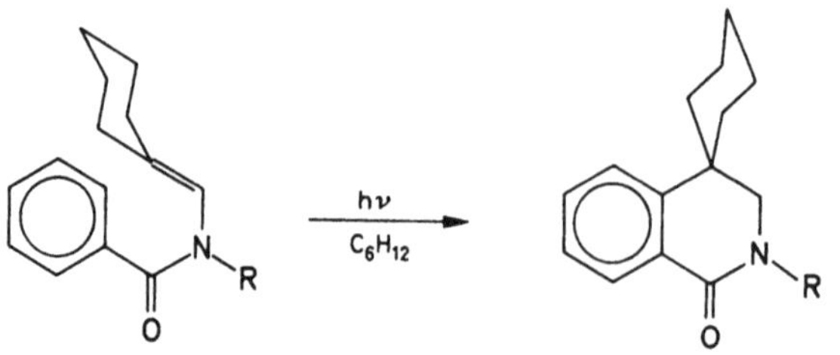

(3.40)

4 Photoaddition Reactions in Synthetic Organic Chemistry

4.1 Hydrogen Abstraction Reactions

In this chapter are summarized the photochemical reactions wherein the primary chemical event is inter- or intramolecular hydrogen transfer to the excited chromophor. In intermolecular reactions hydrogen abstraction usually implies reduction or hydrodimerization of the excited molecule; intramolecular hydrogen abstraction is frequently followed by either ring closure of the diradical or fragmentation to afford unsaturated molecules.

4.1.1 Photoreduction of Molecules with a C—N Double Bond

Molecules containing the imino function are commonly reduced to hydrodimeric products when irradiated in isopropyl alcohol, as illustrated in (4.1)[401] and (4.2)[402]. The photoreduction of nitrogen containing heterocycles has been reviewed recently[403].

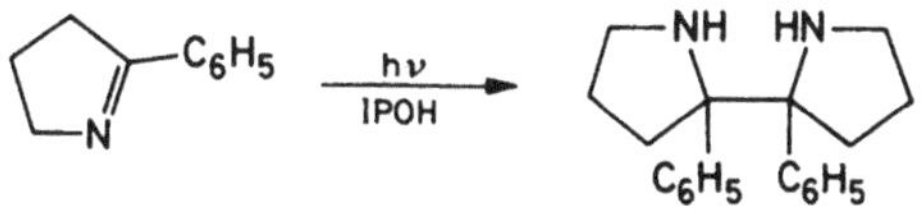

(4.1)

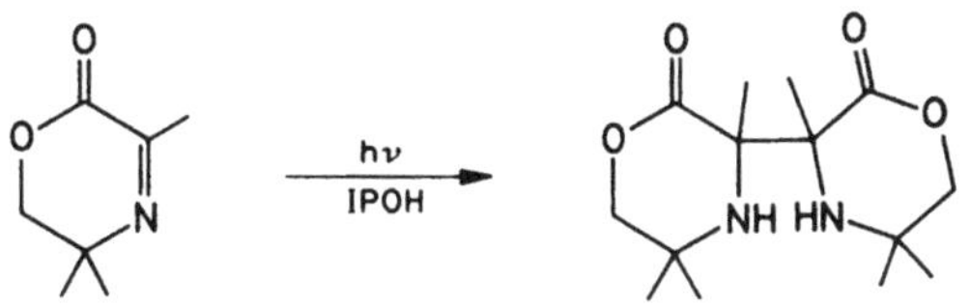

(4.2)

4.1.2 Intramolecular Hydrogen Abstraction by the C—O Double Bond

The process of intramolecular abstraction of an hydrogen atom from the γ carbon atom by the excited carbonyl group is commonly referred to as "Norrish type II" reaction. The resulting diradical can close to a cyclobutanol or cleave to an olefin and an enol, the latter one then tautomerizing to a carbonyl compound (4.3).

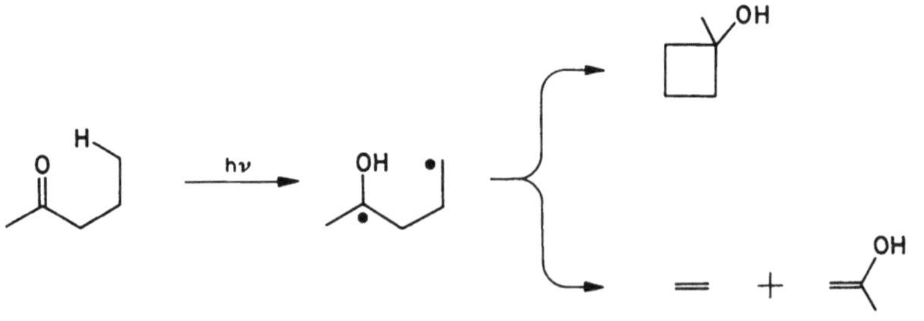

(4.3)

This efficient cyclobutanol forming reaction has been utilized in the synthesis of terpenes[404] (4.4), medium ring compounds[405] (4.5) and strained molecules as in (4.6)[406] and (4.7)[407]. In the last example the tricyclo[2.2.0.0^{2,5}]hexane skeleton is formed.

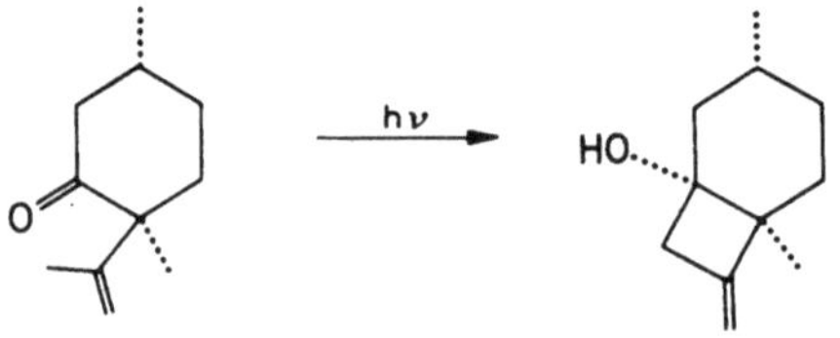

(4.4)

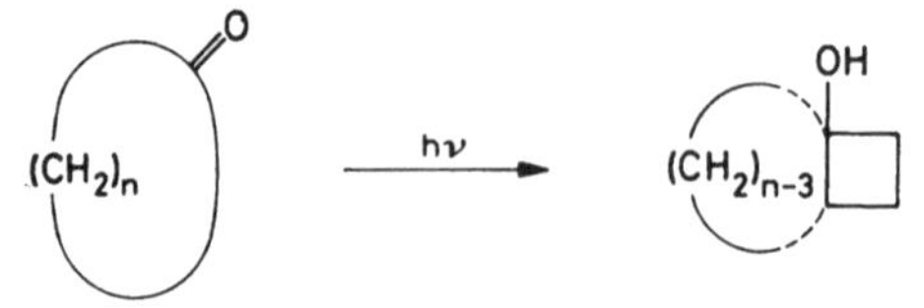

(4.5)

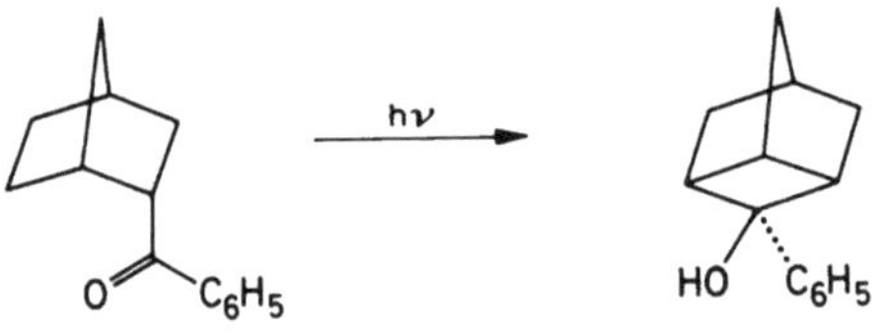

(4.6)

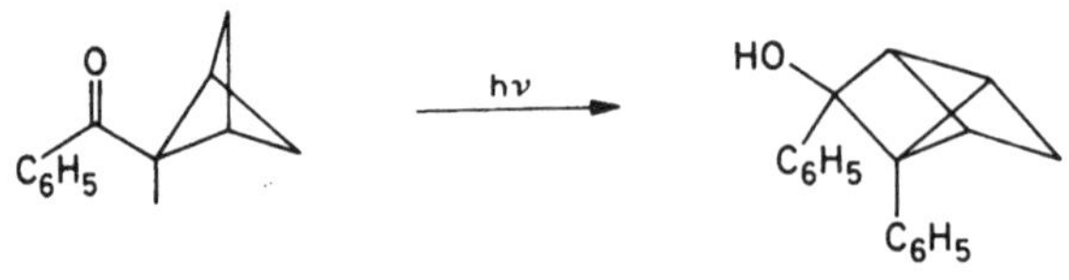

(4.7)

The cleavage of the intermediate 1,4-diradical can also become the major path as in the synthesis of methylenecyclopentane (4.8)[408].

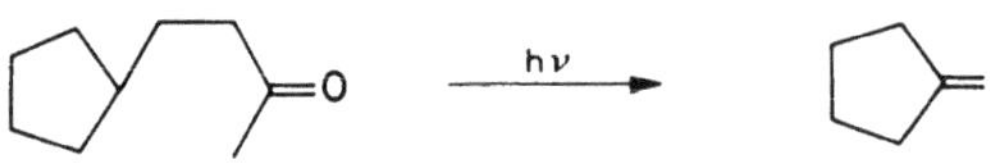

(4.8)

If no γ-H atom is available, or if for steric reasons abstraction of a δ-H atom is facilitated, this latter reaction occurs with formation of a cyclopentanol. A series of such sequences has been used in the synthesis of dodecahedrane [409a, b].

α,β-Unsaturated ketones afford cyclohexanones instead of cyclobutanols as one of the radical centres in the originally formed 1,4-diradical is an allylic moiety. This is illustrated for a 2-cyclopentenone [410] (4.9) and for a butadiene-quinone Diels-Alder adduct (4.10) [411].

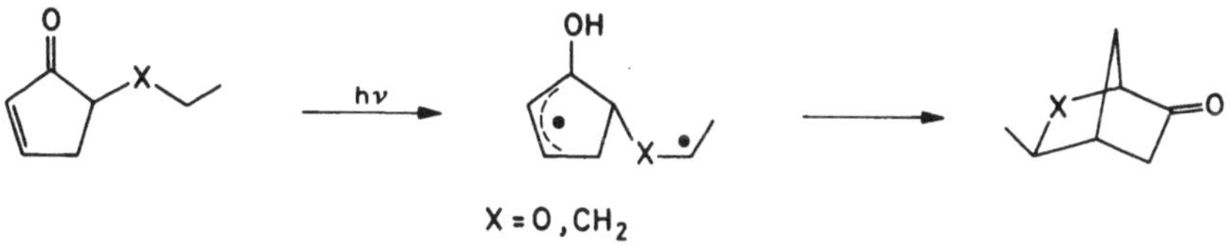

(4.9)

(4.10)

An interesting application of this Norrish type II reaction consists in the photochemical oxydation of alcohols to carbonyl compounds by irradiation of their pyruvic esters (4.11) [412].

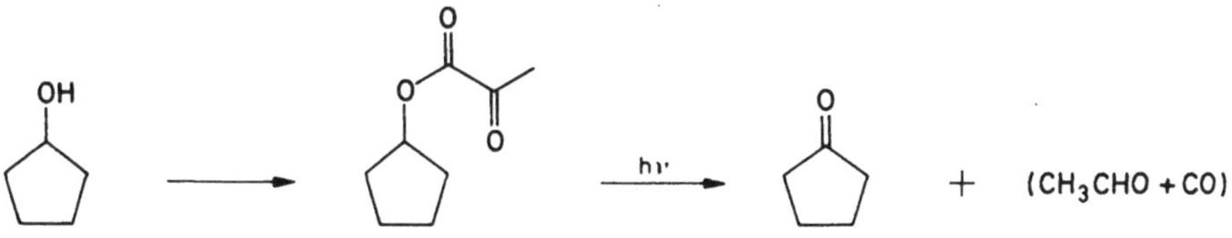

(4.11)

The irradiation of ortho tolyl ketones affords dienols [413] *via* an analogous intramolecular H-abstraction. The dienols formed usually tautomerize back to starting material, but they can also be trapped by dienophiles in inter- [414a] (4.12) and intramolecular [414b] (4.13) Diels-Alder reactions. This latter sequence has been successfully utilized in a stereoselective synthesis of estrone.

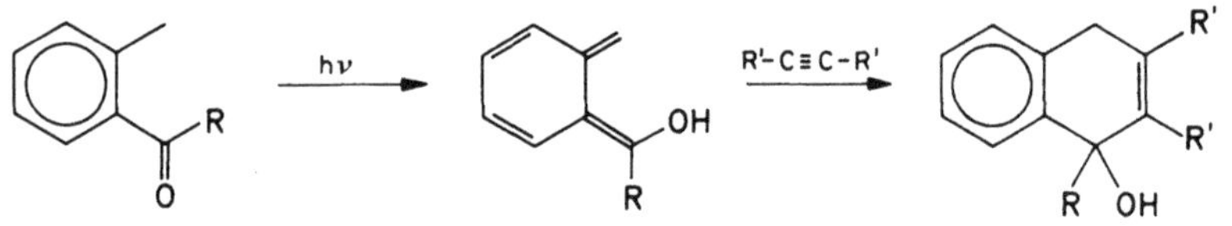

(4.12)

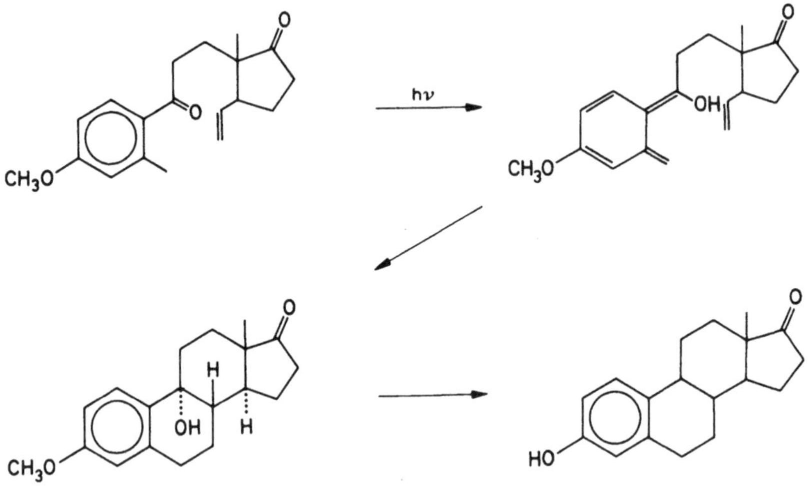

(4.13)

A dienol is also formed *via* hydrogen abstraction by the excited carbonyl group from a second enone molecule in (4.14). This dienol tautomerizes in C_6F_6 to give the β,γ-unsaturated ketone selectively, the overall reaction thus being deconjugation of the α,β-unsaturated ketone [415].

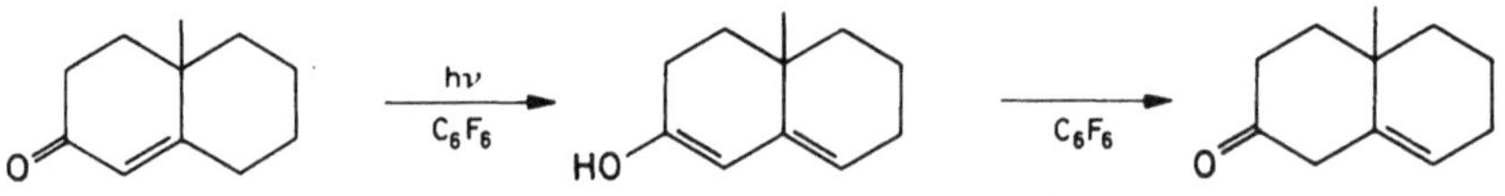

(4.14)

Phthalimides undergo a variety of photochemical reactions [416]. The sequence of a Norrish type II reaction has been utilized in the synthesis of medium ring thiacyclols (4.15) [417].

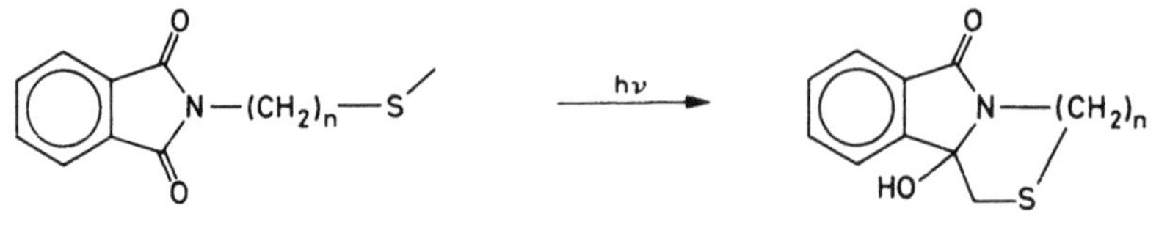

(4.15)

4.1.3 Photoreduction of Carbonyl Compounds

The reduction of carbonyl compounds is most often better achieved by electrochemical methods or by using conventional hydride or hydrogenation reactions. Nevertheless the formation of benzpinacol from benzophenone, although now known for more than eighty years, is still a matter of study (4.16)[418].

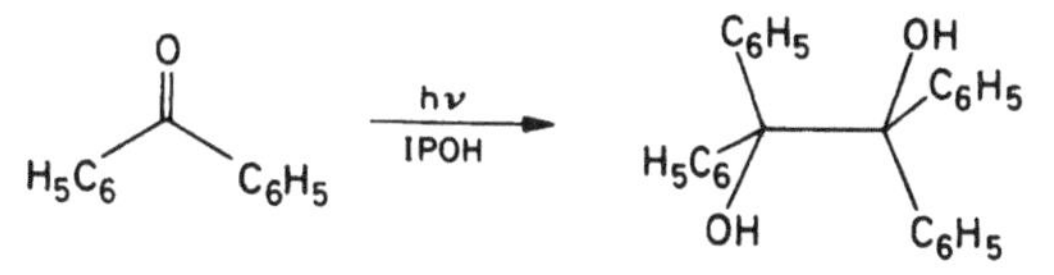

(4.16)

Allylic hydrogens can easily be abstracted by excited carbonyl compounds. Recombination of the two radicals afford products which are apparently formed by RH addition to the carbonyl group (4.17)[419].

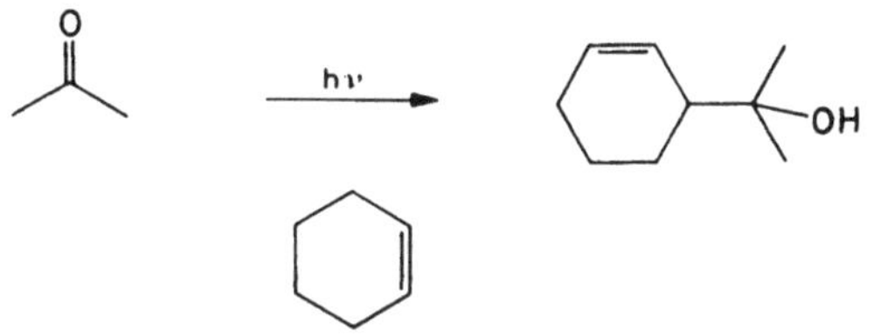

(4.17)

Conformationally constrained 2-cyclohexenones which cannot undergo the lumiketone rearrangement (cf. chapter 3.1.5) are efficiently reduced to cyclohexanones (4.18)[420].

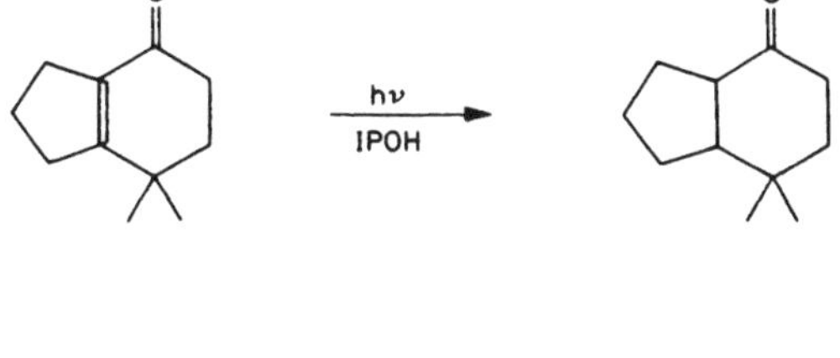

(4.18)

4.1.4 Hydrogen Abstraction by the C—C Double Bond of an α,β-Unsaturated Ketone

As discussed before, α,β-unsaturated ketones are usually reduced to saturated ketones in isopropyl alcohol as solvent, the reaction proceeding *via* the enol after H-abstraction by the carbonyl group. When some selected α,β-enones are irradiated in their π-π* absorption band (i.e. excitation to the S_2-state) products are formed which arise formally *via* hydrogen abstraction by the α-carbon of the enone. This reaction has been utilized in the synthesis of propellanes (4.19)[421]. For the reaction to occur, the substituent on $C_{4\alpha}$ of the naphthalenone system must be able to give a resonance stabilized radical when losing the hydrogen atom ($CH(OCH_3)_2$, $CH{=}CHCH_3$); otherwise 1,3-migration of the substituent becomes the major reaction.

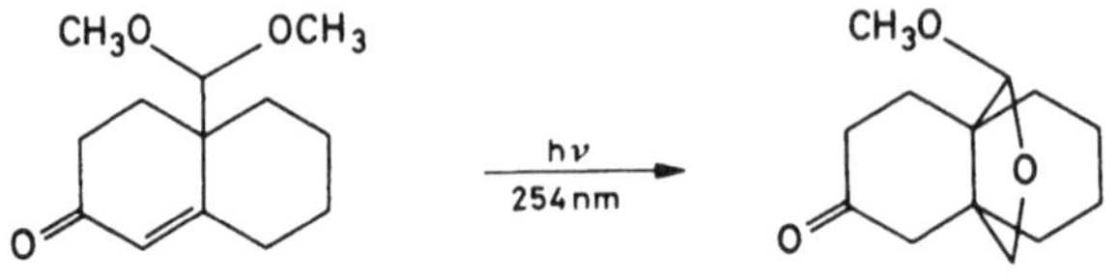

(4.19)

4.2 Photodimerization Reactions

In this type of process an excited molecule adds to a second — identical — molecule in its ground state, usually with formation of a ring. These dimerizations are thus most commonly intermolecular reactions, but obviously the two reactive moieties can also be linked together, e.g. by an alkyl chain. Such intramolecular photodimerization reactions have been studied in detail[422].

4.2.1 Cyclobutane-Forming Photodimerizations

Simple olefins can photodimerize to cyclobutanes in the presence of cuprous salts as sensitizers (4.20)[423]. Copper (I) trifluoromethanesulfonate has been utilized successfully (4.21)[424].

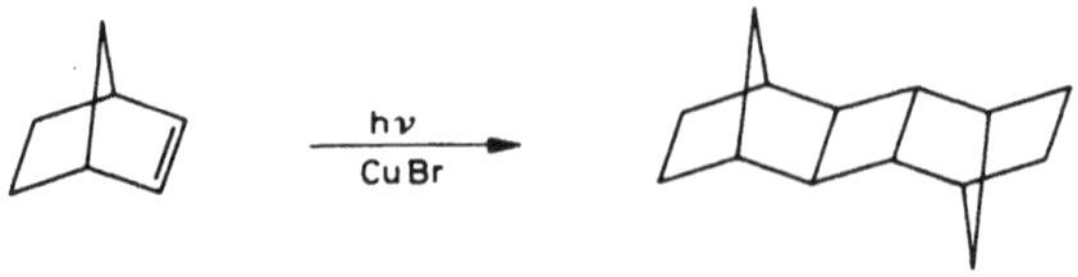

(4.20)

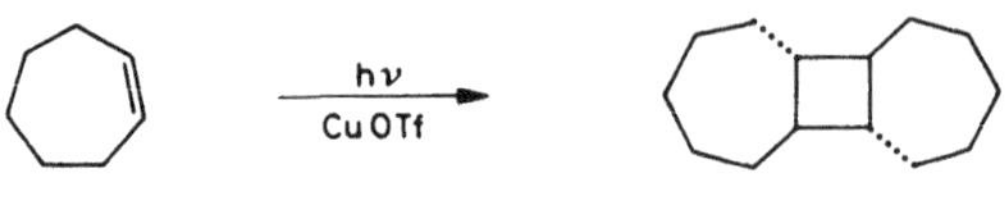

(4.21)

Stilbene-like olefins can be irradiated directly to afford cyclodimers (4.22)[425].

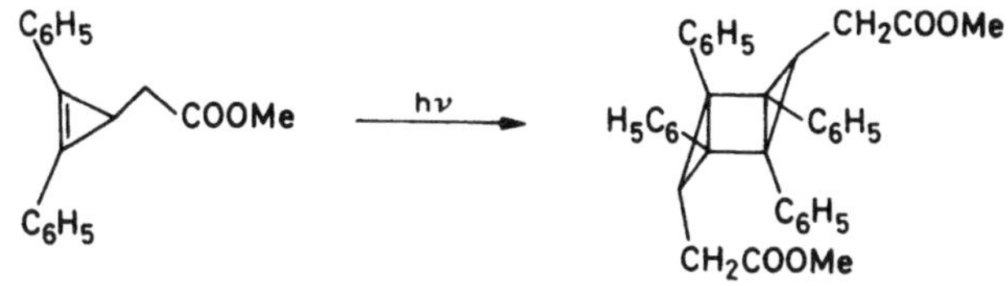

(4.22)

Cyclic α,β-enones photodimerize *via* the T_1-state, as in (4.23)[426] and (4.24)[427]. By chosing the appropriate substituents the reactions proceed regio- and stereoselectively. In contrast, pyrimidine bases dimerize from the first excited singlet state to afford head to tail products (4.25)[428].

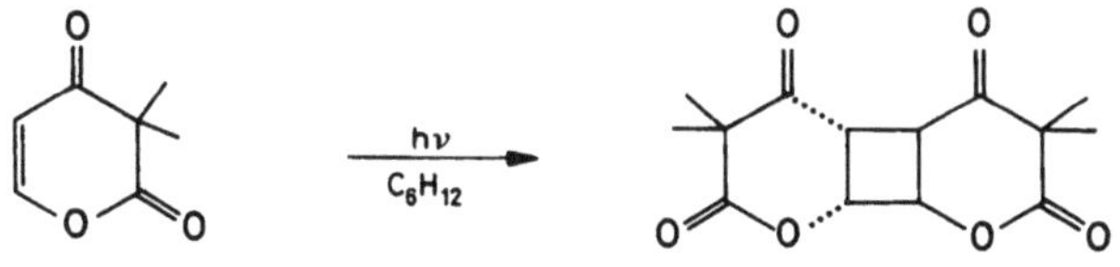

(4.23)

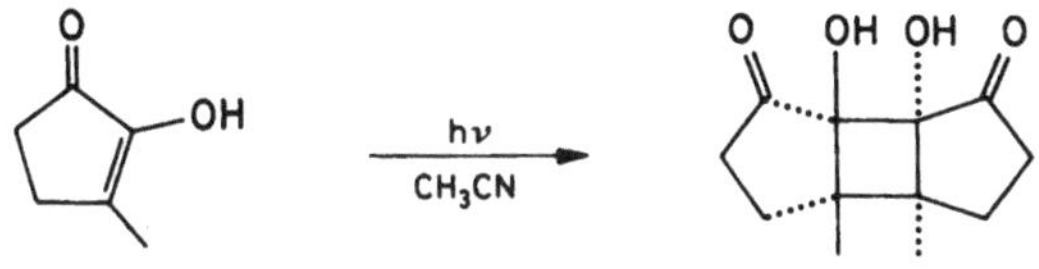

(4.24)

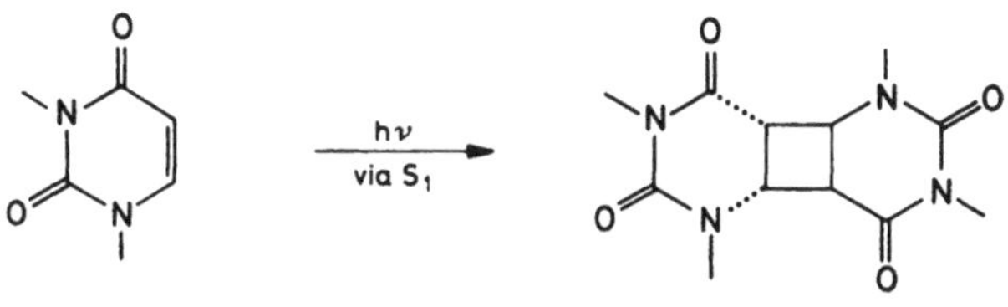

(4.25)

4.2.2 Photodimerization of Arenes

The [4+4]photodimerization of anthracene has already been reported more than 115 years ago (4.26)[429].

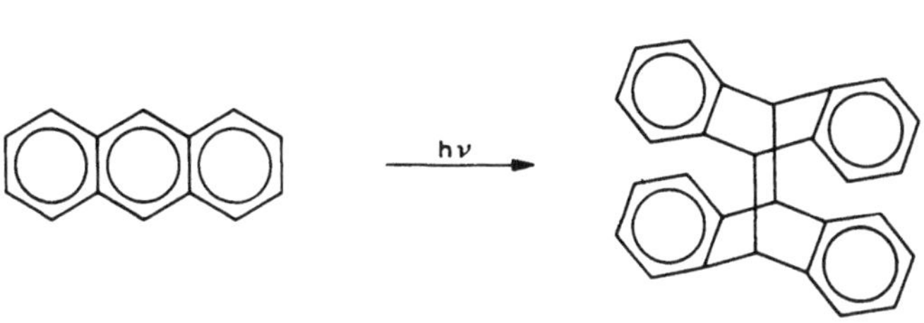

(4.26)

Similar formations of bisanthracenes have been studied with a variety of substituents. An interesting example of a cation-assisted intramolecular anthracene dimerization where the bisanthracene formation includes cyclization of a polyether chain to a crown ether is illustrated in (4.27). In the absence of Li^+ the "dimer" reverts back to the open chein compound, but in the presence of Li^+ the crown ether is stabilized so that the product itself also becomes more stable [430].

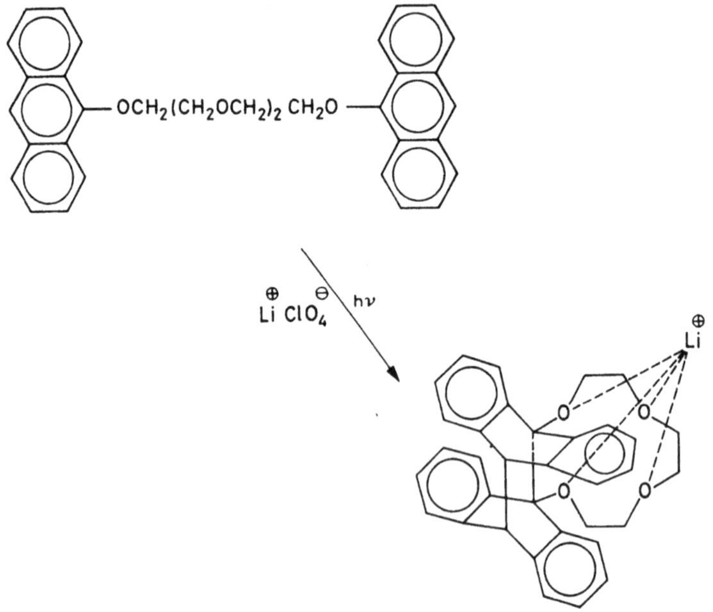

(4.27)

Analogous [4+4]photodimers are obtained in irradiating naphthalene derivatives (4.28) [431].

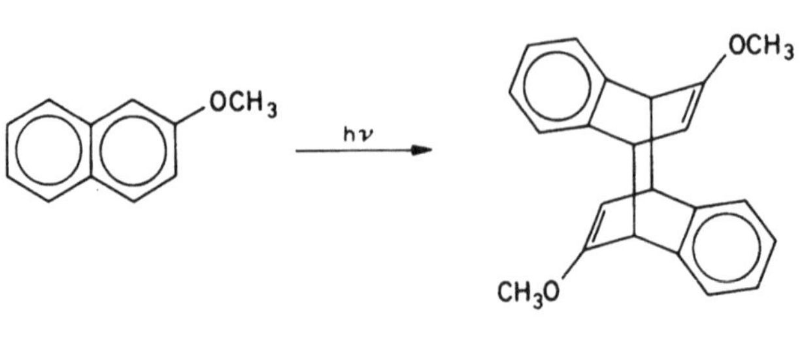

(4.28)

While benzene and substituted benzenes usually do not dimerize, pyridine derivatives undergo [4+4]photodimerizations [432] (4.29).

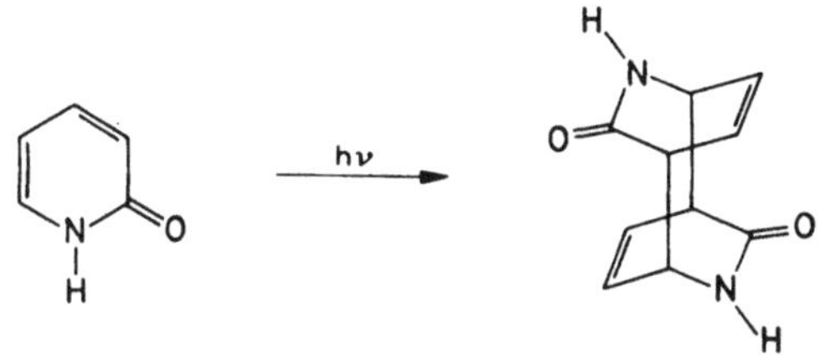

(4.29)

4.2.3 Photodimerization of Imines

Simple aliphatic imines are often considered as being photounreactive. One of the few reported examples of a 1,3-diazetidine formation concerns an imine of difluoroacetone (4.30) [433].

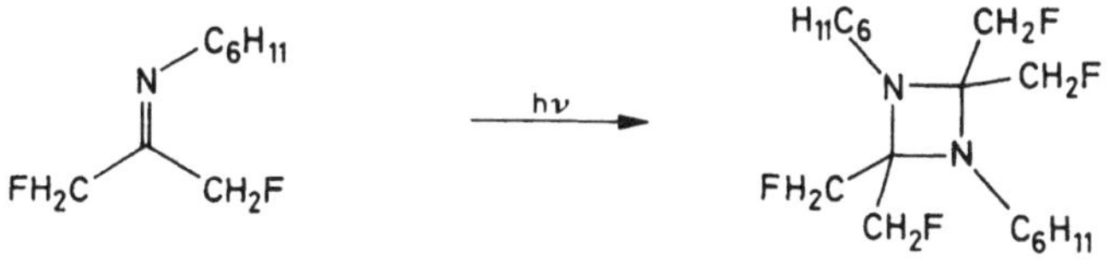

(4.30)

4.3 Photocycloadditions to the C—C Double Bond

This chapter deals with [2+2]cycloadditions of various chromophors to an olefinic double bond with formation of a four-membered ring, with reactions proceeding as well in an intermolecular as in an intramolecular pattern. Due to the variety of the starting materials available (ketones, enones, olefins, imines, thioketones, etc. ...), due to the diversity of products obtained, and last but not least, due to the fact that cyclobutanes and oxetanes are not accessible by such a simple one-step transformation in a non-photochemical reaction, the [2+2]photocycloaddition has become equivalent to the (thermal) Diels-Alder reaction in importance as for ring construction in organic synthesis.

4.3.1 Intramolecular Photocycloaddition of two C—C Double Bonds

As already mentioned in discussing the photodimerization of alkenes (cf. chapter 4.2.1), direct irradiation of the olefinic double bond is most often inefficient from the practical point of view as olefins do not absorb light of wavelength longer than 200–210 nm and because the rate of intersystem crossing from S_1 to T_1 is small. Triplet sensitizers are therefore used in these reactions; these are either acetone, which then also serves as solvent, or the irradiations are run in the presence of small amounts of cuprous salts. Selected examples of such reactions are the synthesis of vinylbicyclo-[2.1.1]hexane [434] (4.31) or of a bishomocubane (4.32) [435]. The conversion of norbornadiene to quadricyclane [436] (4.33) represents a prototype reaction for chemical storage of solar energy; properly substituted norbornadienes absorb visible light to afford the quadricyclanes in high quantum yields. The quadricyclanes revert catalytically to the norbornadienes in a strongly exothermic reaction. In such a cycle light is thus transformed into heat.

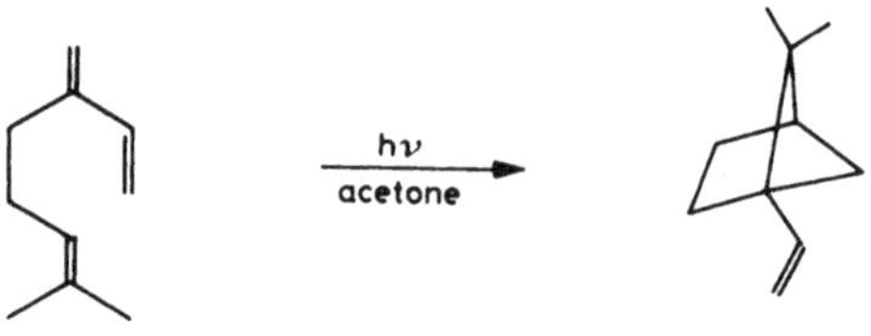

(4.31)

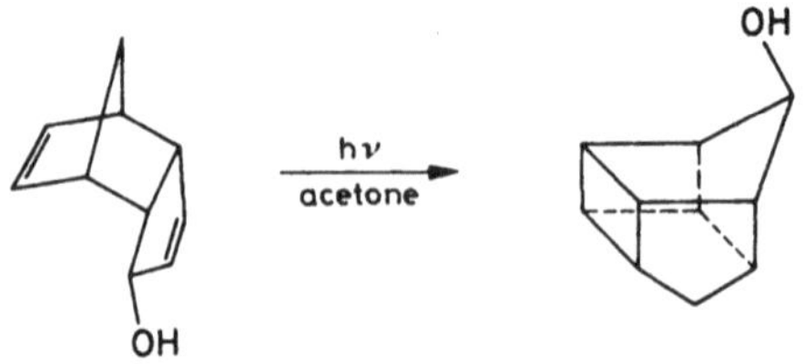

(4.32)

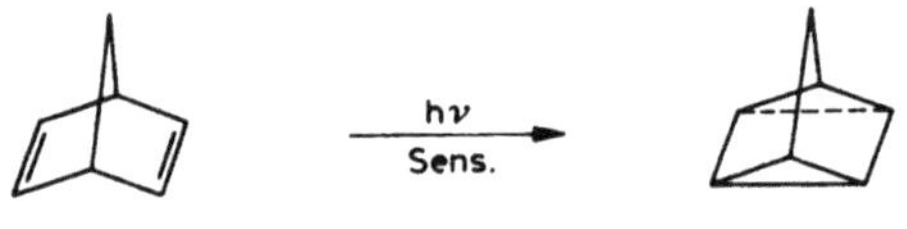

(4.33)

The irradiation of 1,5-cyclooctadienes can lead to either linear [437] (4.34) or crossed [438] (4.35) adducts as demonstrated in the examples below.

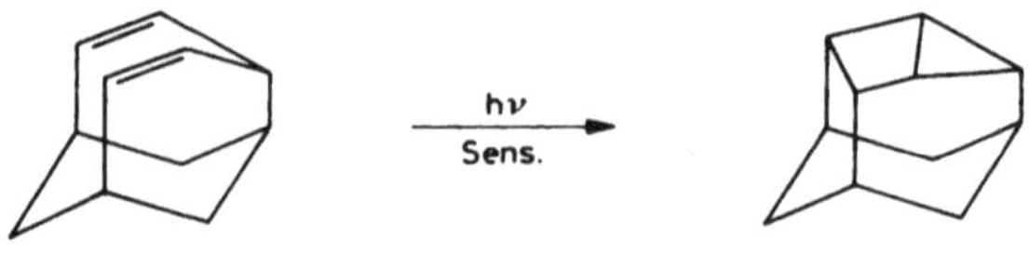

(4.34)

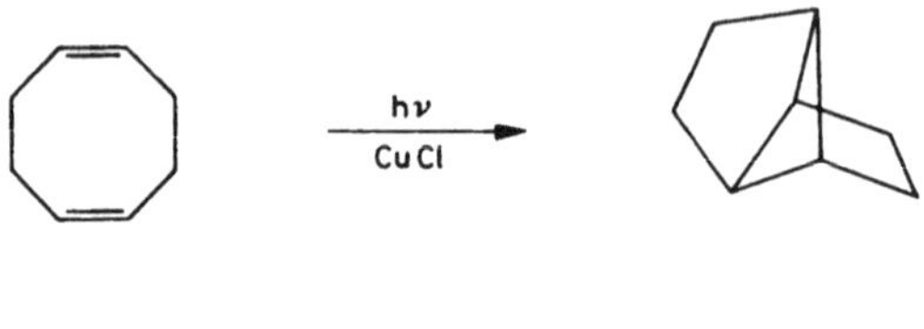

(4.35)

Acyclic dienes as 1,6-heptadienes can be converted to bicyclic compounds, e.g. bicyclo[3.2.0]heptanes, in the presence of cuprous triflate (4.36)[439].

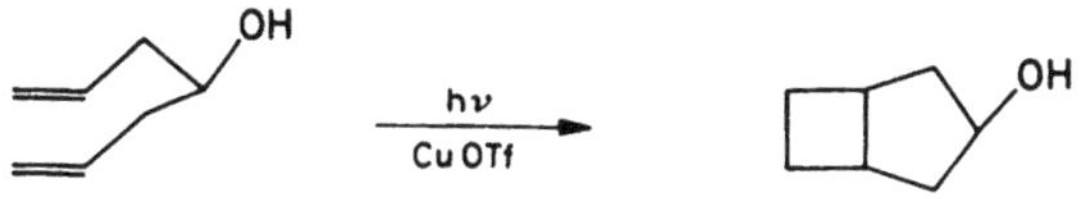

(4.36)

4.3.2 Photocycloaddition of Arenes to Olefins

The synthetic applications[440] and mechanistic aspects[441] of intermolecular photocycloaddition reactions of arenes to olefins have been reviewed recently. Intramolecular cycloadditions[442a, b] have been studied in the context of the photochemical behaviour of bichromophoric molecules, as to investigate interchromophoric interactions in polyfunctional molecules. Three types of addition products can be formed in the photocycloaddition of benzene to an alkene (4.37)[441].

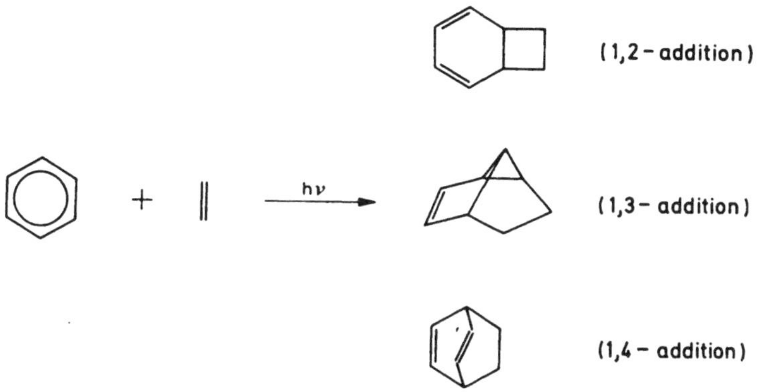

(4.37)

1,2-Additions of benzene, not to olefins but to acetylenes have been utilized in the synthesis of cyclooctatetraenes (4.38)[443].

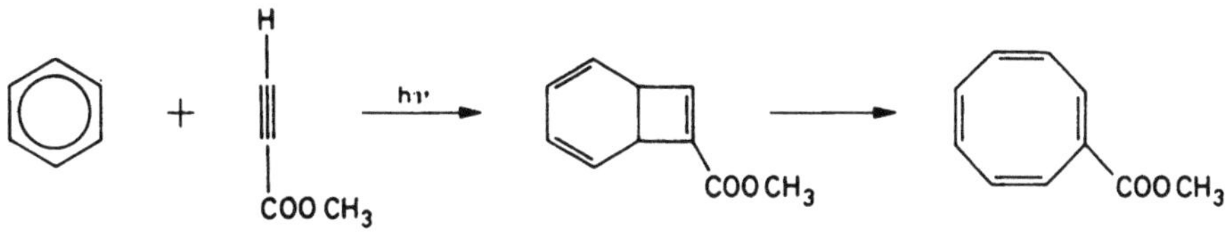

(4.38)

Naphthalene and substituted naphthalenes add to olefins[444] (4.39) and to acetylenes[445] (4.40) to give 1,2-adducts. In the latter case the primary addition product undergoes a further [2+2]intramolecular cycloaddition.

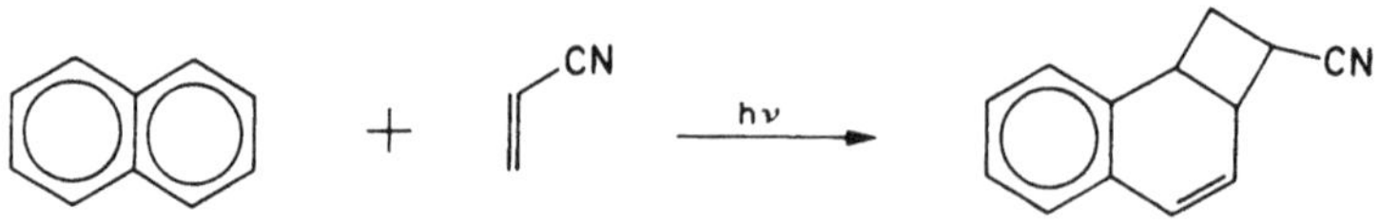

(4.39)

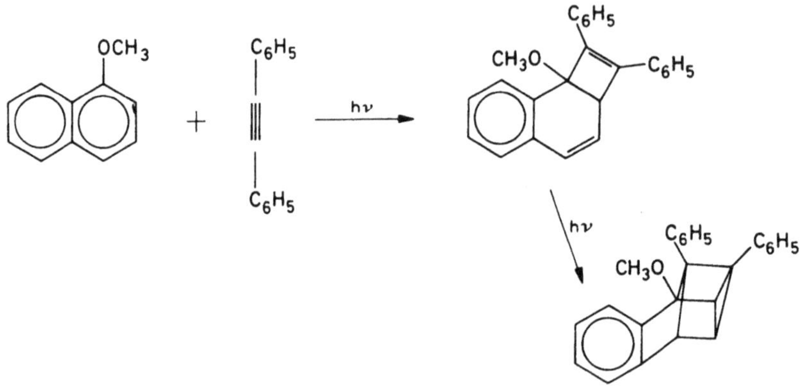

(4.40)

The addition of anisole to cyclopentene gives an 1,3-adduct which rearranges to tricycloundecenone in the presence of traces of acid (4.41)[446].

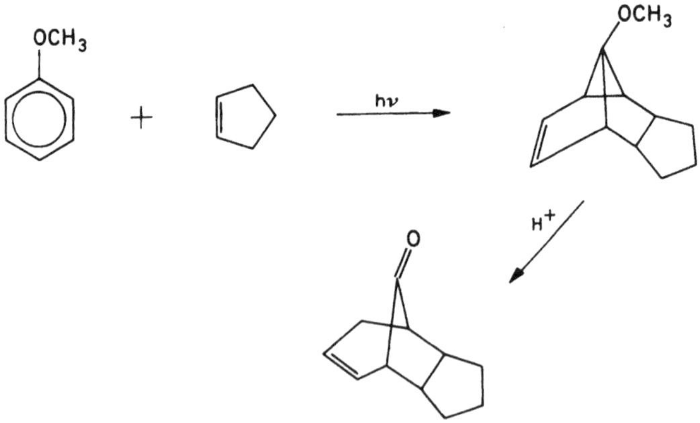

(4.41)

The uniqueness of the intramolecular 1,3-photocycloaddition of arenes to olefins with the developement of three new rings and up to six stereocentres has been efficiently demonstrated in the synthesis of (±)-α-cedrene from the arenealkene in (4.42)[447]. Both primary photoadducts are converted to the same ketone which is then reduced to the sesquiterpene.

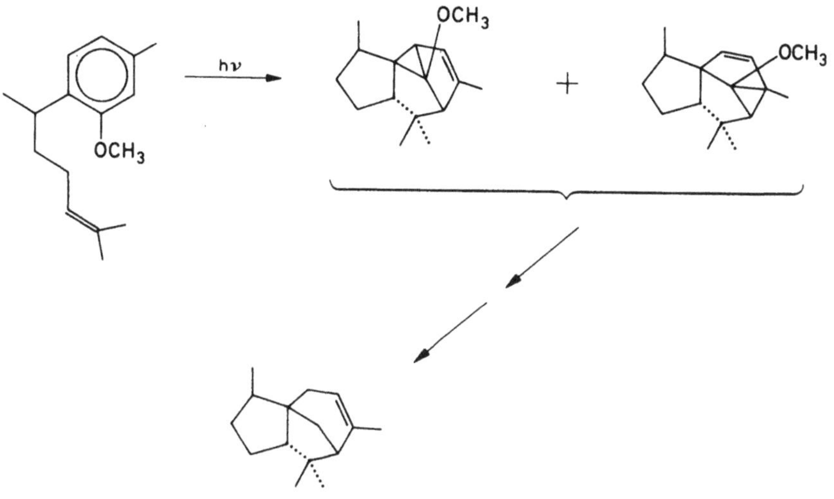

(4.42)

Addition products of the 1,4-type are only formed in some special examples. In the addition of benzene to allene the bicyclo[2.2.2]octane is the major product (4.43)[448].

(4.43)

4.3.3 Photocycloaddition of α,β-Unsaturated Ketones to Olefins

Among the photoreactions affording four-membered rings, the cycloaddition of α,β-unsaturated ketones to alkenes is by far the most widely utilized reaction of this type[449]. The advantages of an enone over a simple olefin as photoreactant are quite obvious: first, enones absorb light at longer wavelengths making it easier to work with conventional light sources; second, intersystem crossing from S_1 to T_1 is efficient for enones, so that no sensitizer is needed, in contrast to simple alkenes; and last but not least by this way the photoproduct is an acylcyclobutane vs. a nonfunctionalized ring, thus increasing the possibilities for further synthetic transformations.

The enone system itself is usually part of a five- or six membered ring, although acyclic α,β-unsaturated ketones and enols of β-diketones are also found to undergo cycloadditions under certain conditions. For seven- and higher membered rings the primary photochemical event is Z—E isomerization around the C—C double bond, the E-isomer then eventually undergoing further thermal reactions.

Mechanistic evidence indicates [450,451] that the triplet enone first approaches the olefinic partner to form an exciplex. The next step consists in the formation of one of the new C—C bonds to give a 1,4-diradical, which is now the immediate precursor of the cyclobutane. Both exciplex and 1,4-diradical can decay resp. disproportionate to afford ground state enone and alkene. Eventually oxetane formation, i.e. addition of the carbonyl group of the enone to an olefin is also observed [452]. Although at first view the photocycloaddition of an enone to an alkene would be expected to afford a variety of structurally related products, the knowledge of the influence of substituents on the stereochemical outcome of the reaction allows the selective synthesis of the desired annelation product in intermolecular reactions [453,454a, b]. As for intramolecular reactions, the substituent effects are made up by structural limitations [449].

Acyclic hexadienones have been found to undergo intramolecular cyclization to bicyclohexanones [455,456] (4.44).

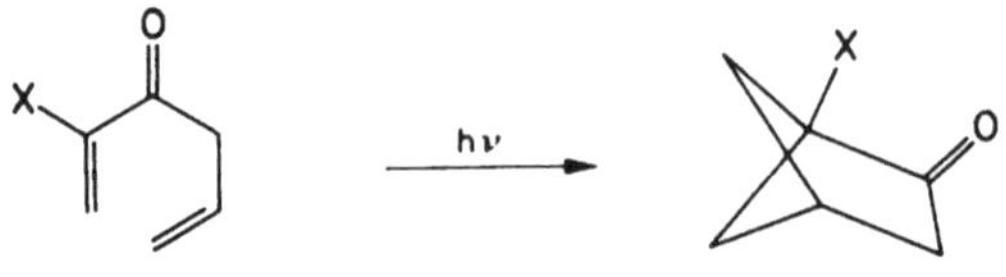

(4.44)

Cyclopentenones with a propenyl side chain on C_4 or C_5 give tricyclooctanones on irradiation, as illustrated in (4.45) [457] and (4.46) [458].

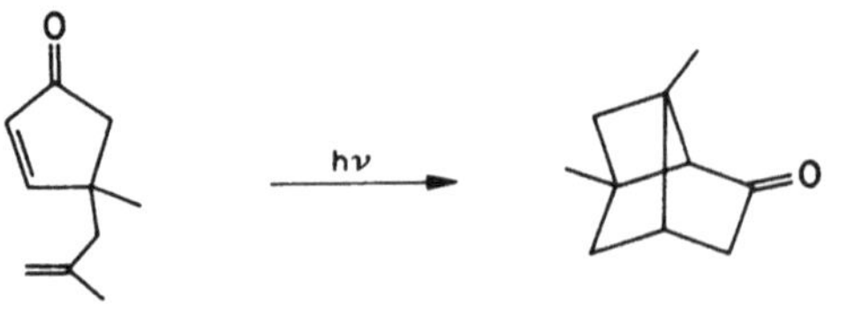

(4.45)

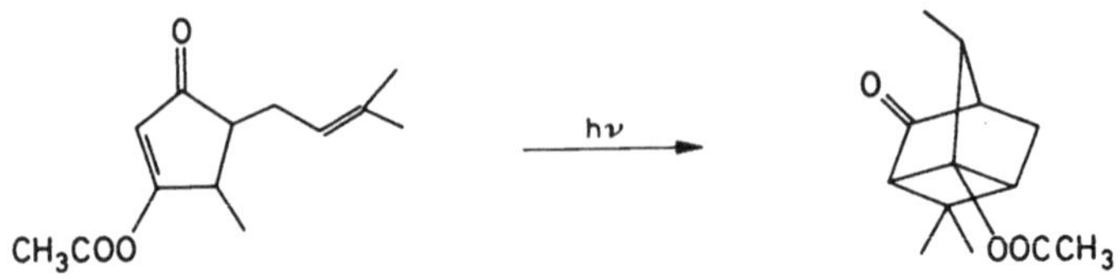

(4.46)

Under similar conditions 6-allyl-2-cyclohexenones cyclize to tricyclononanones [459] (4.47).

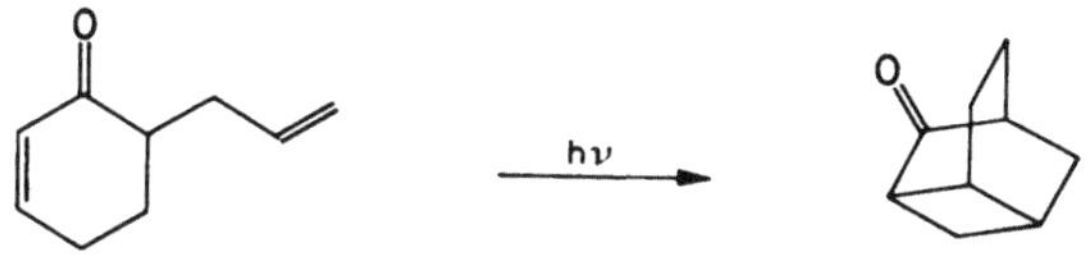

(4.47)

The first claimed synthesis of a fenestrane precursor follows a similar route [460a] (4.48). In general such intramolecular cycloadditions represent an efficient way of synthesizing centropolycyclanes [460b].

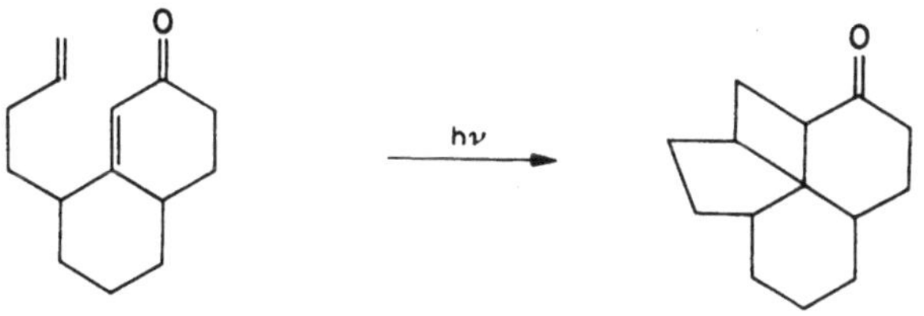

(4.48)

Such reactions have also often been used in the synthesis of sesquiterpenes as (±)-epiprecapnelladiene [461] (4.49), (±)-isocomene [462] (4.50) or of hirsutane derivatives (4.51) [463].

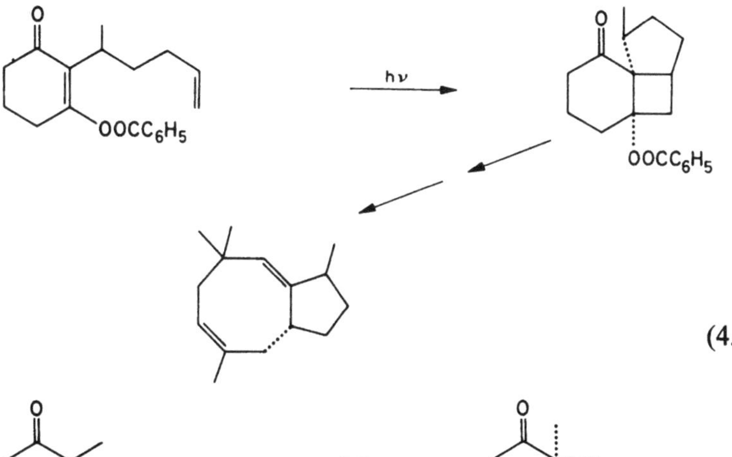

(4.49)

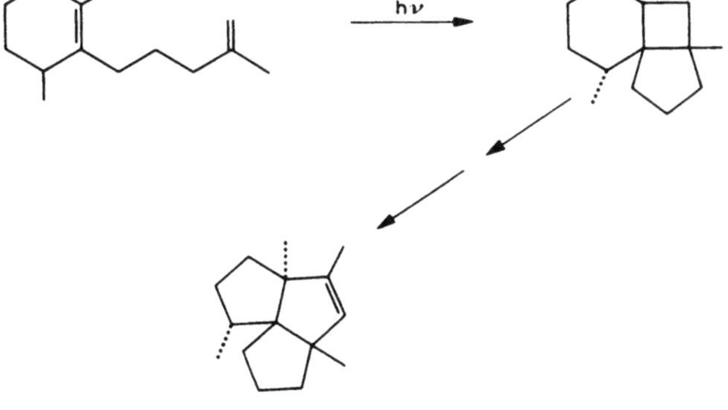

(4.50)

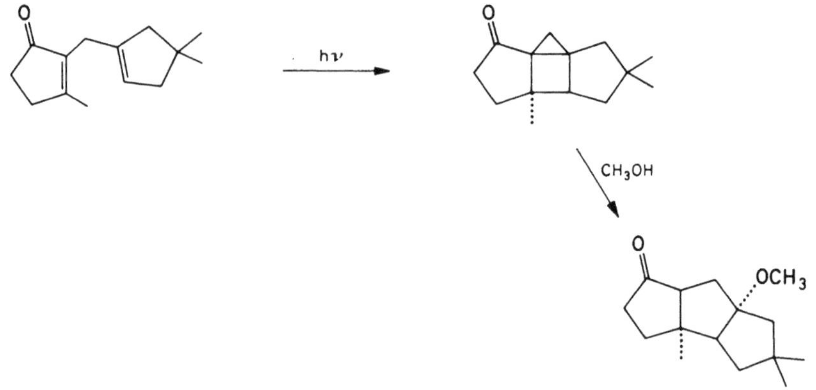

(4.51)

The intramolecular addition of an enone to an *in situ* generated ketene has been utilized in the synthesis of a tricyclodecanedione derivative (4.52) [464].

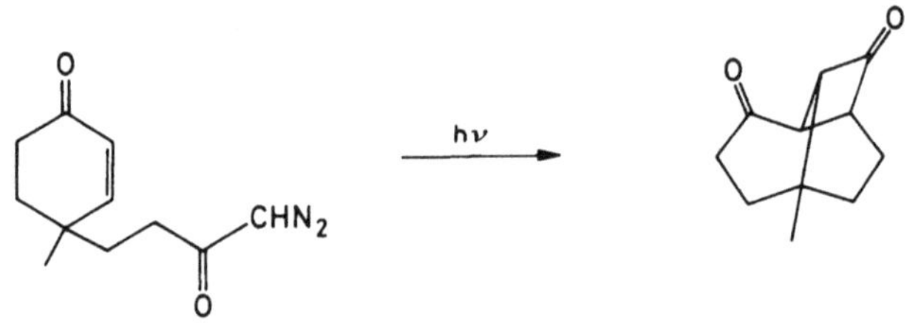

(4.52)

Bicycloheptanones formed by intermolecular cycloaddition of 2-cyclopentenones to olefins, as in (4.53) [465], are valuable synthons in organic chemistry [454a)].

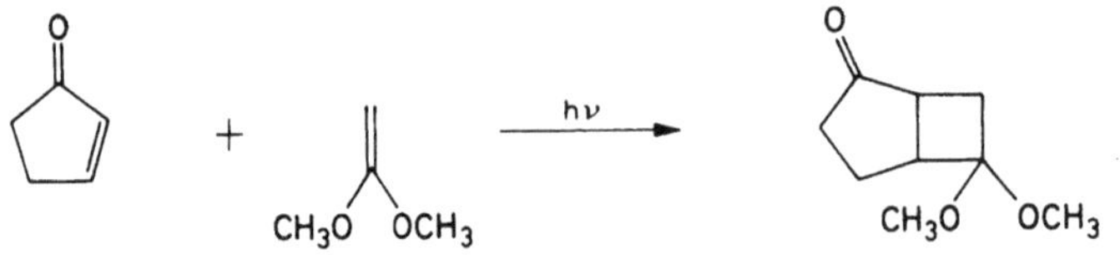

(4.53)

The sesquiterpenes (±)-10-epijuneol [466] (4.54) and (±)-calameon [467] (4.55) have been obtained *via* addition of 6-isopropyl-3-methyl-2-cyclohexenone to cyclobutene derivatives. A similar sequence has been applied for the synthesis of an eudesmane precursor (4.56) [468].

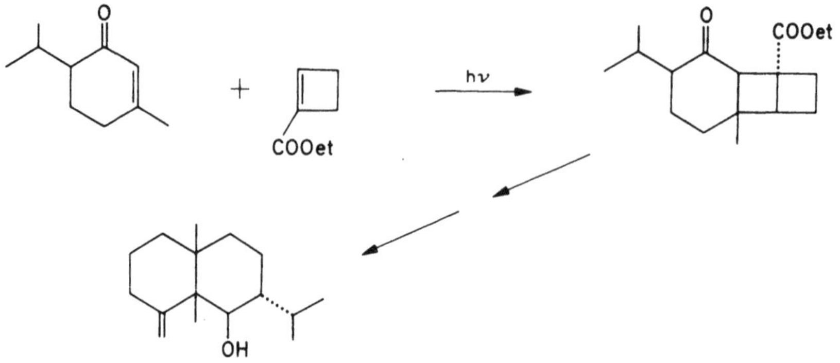

(4.54)

(4.55)

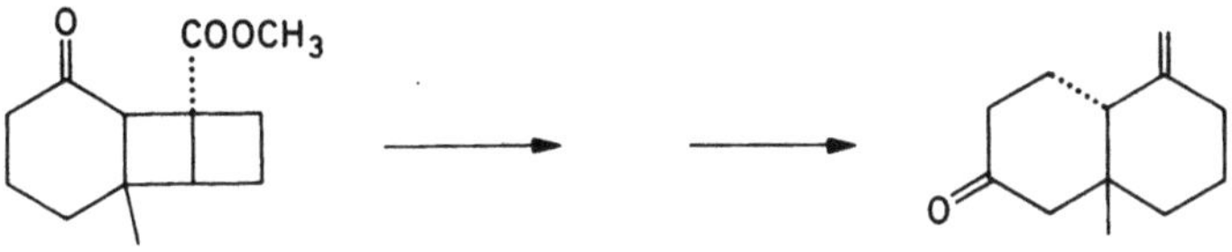

(4.56)

The addition of the 2-substituted 2-cyclopentenone in (4.57) to ethylene affords an intermediate which is converted to tricyclo[4.2.0.0^{1,4}]octane [469a)]. A similar "broken window" compound was also obtained by an intramolecular photocycloaddition of a 2-cyclopentenone (4.58) [469b)].

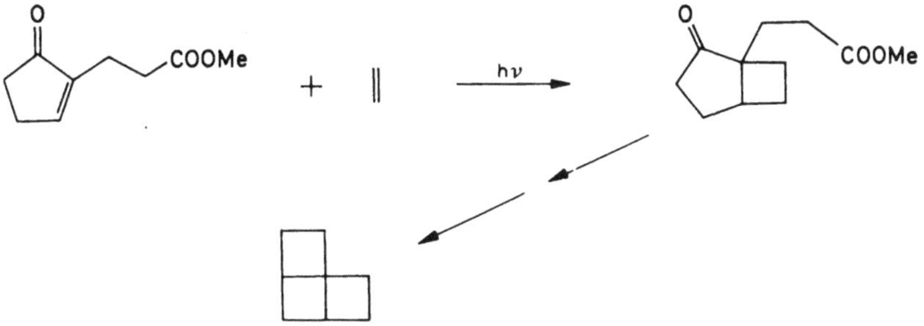

(4.57)

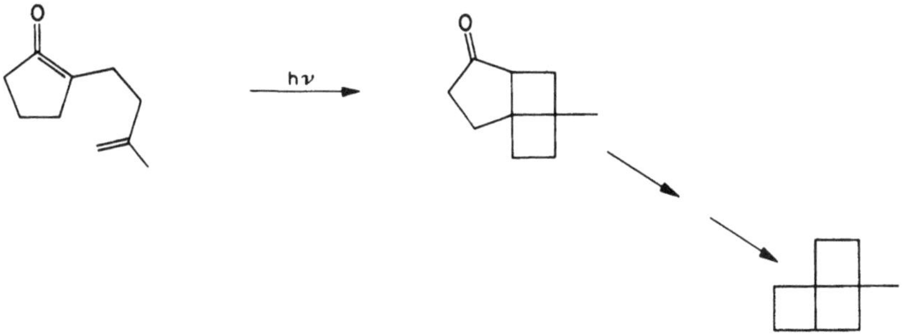

(4.58)

An useful intermediate for the preparation of 1,4-cyclohexanediones is obtained by cycloaddition of 6-isopropyl-3-methyl-2-cyclohexenone to a cyclobutanediol derivative (4.59) [470)].

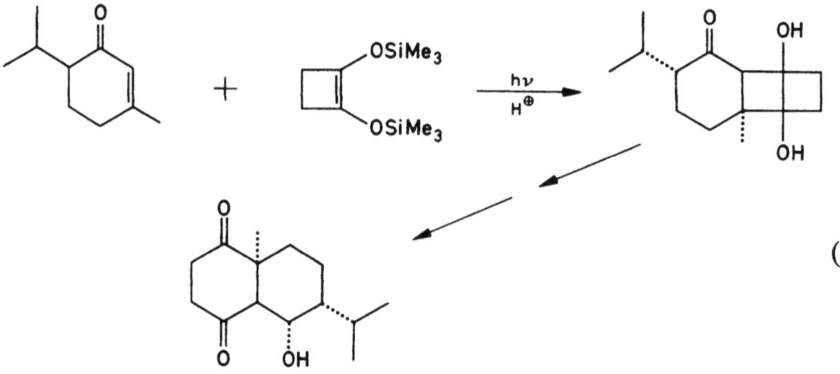

(4.59)

The sesquiterpenes (±)-protoilludene [471] (4.60) and (±)-modhepene [472] (4.61) have also been synthesized via intermolecular cycloaddition reactions.

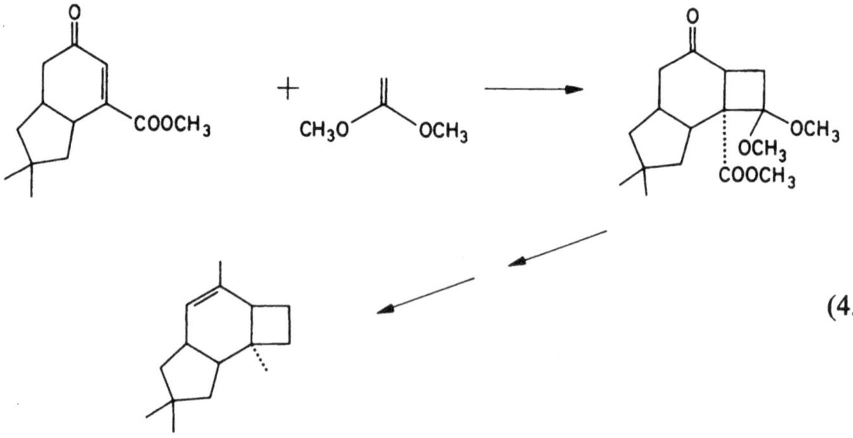

(4.60)

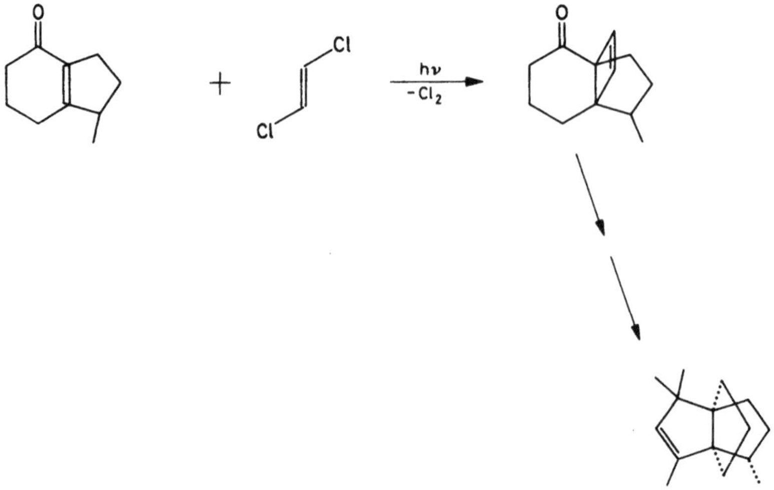

(4.61)

3(2H)-furanones undergo efficient cycloadditions as "oxa-enones" [450]. The cycloadducts have been successfully utilized as synthetic precursors for δ-valerolactones [473] (4.62) or for 2-cyclohexenones (4.63) [474].

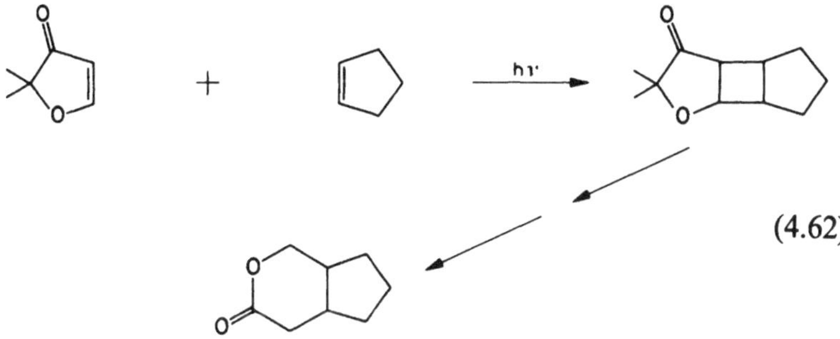

(4.62)

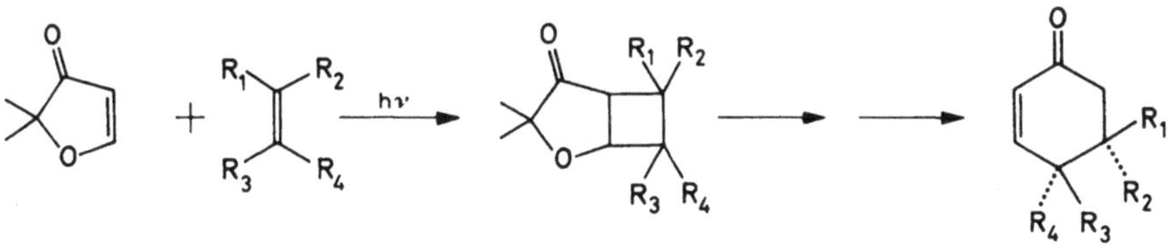

(4.63)

With a similar sequence it has been established that the assumed structural formula for genipic acid is not the correct one [475], as the cyclopentene derivative in (4.64) does not behave like the natural product itself.

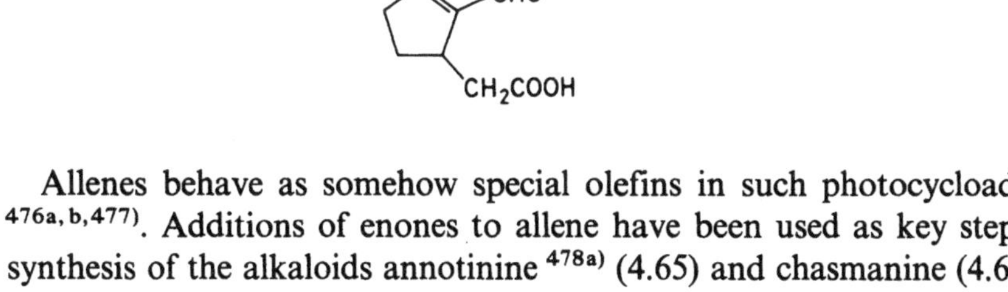

(4.64)

Allenes behave as somehow special olefins in such photocycloadditions [476a, b, 477]. Additions of enones to allene have been used as key step in the synthesis of the alkaloids annotinine [478a] (4.65) and chasmanine (4.66) [478b].

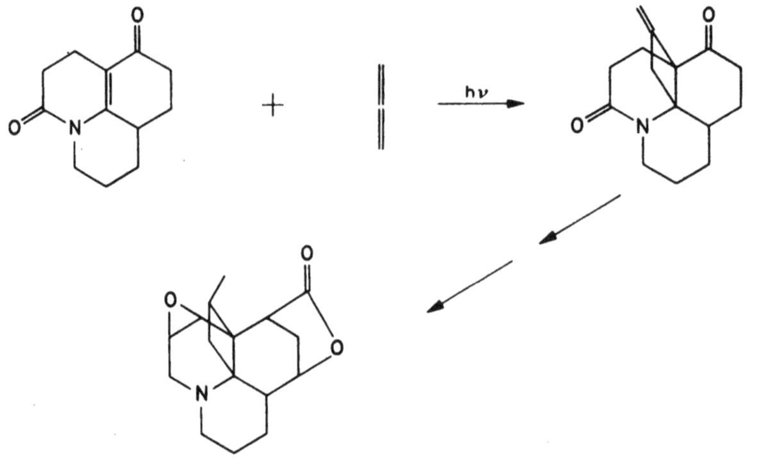

(4.65)

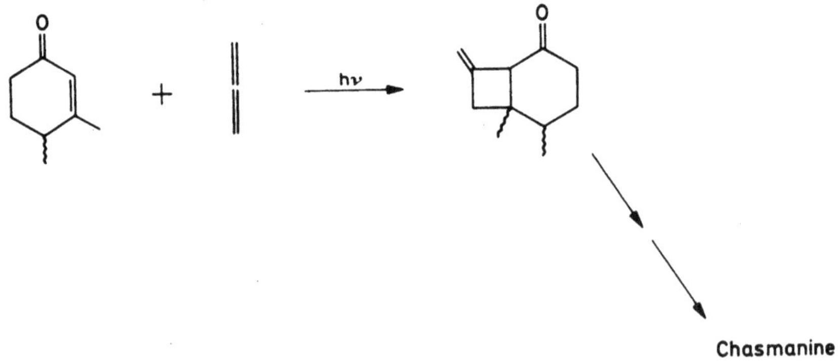

(4.66)

Dienones can behave as enones in additions to dienes [479a)]. The interesting "ladder" compound in (4.67) has been isolated in moderate yields [479b)].

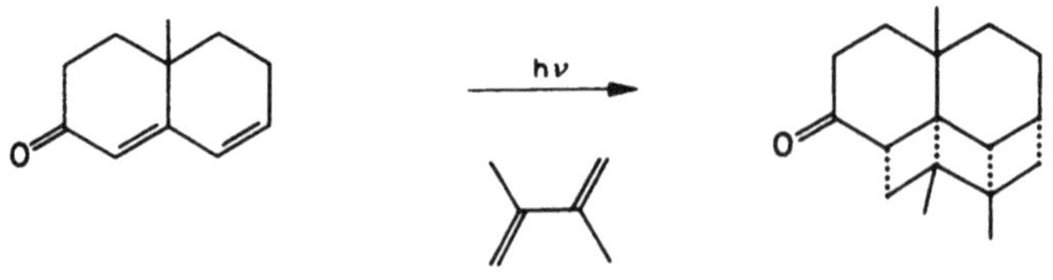

(4.67)

4.3.4 Photocycloaddition of α,β-Unsaturated Acid Derivatives to Alkenes

Like α,β-unsaturated ketones, α,β-unsaturated carboxylic acid derivatives, e.g. lactones and anhydrides, undergo cycloadditions to alkenes. As for the preparative conditions (direct irradiation or sensitized experiments) these compounds are situated somewhere in between enones on the one side and olefins on the other.

Intramolecular cycloadditions of α,β-unsaturated lactones to alkenes have been used as key step in the synthesis of a precursor of the terpene reserpine [480] (4.68) and of fused cyclooctanes (4.69) [481].

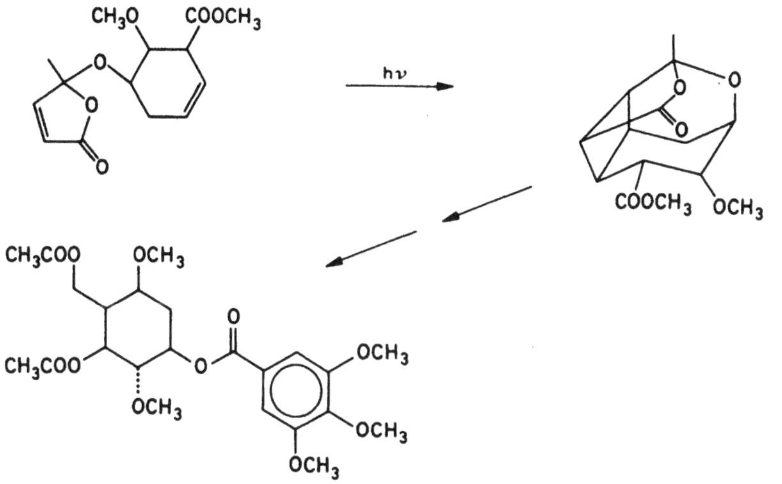

(4.68)

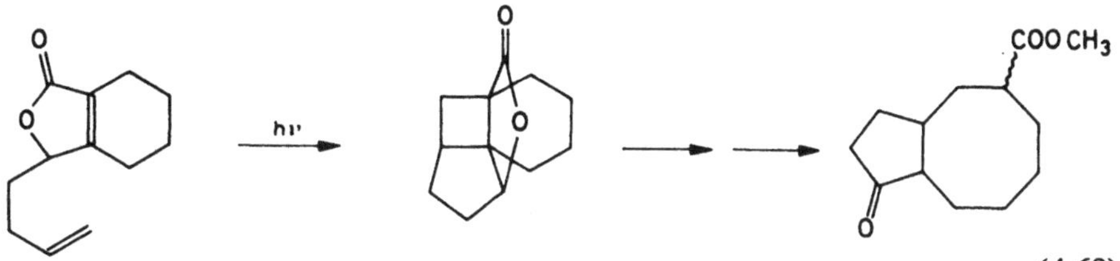

(4.69)

Maleic anhydride adds to ethylene or to acetylene to afford the corresponding cyclobutanes (or cyclobutenes) (4.70) [482a, b].

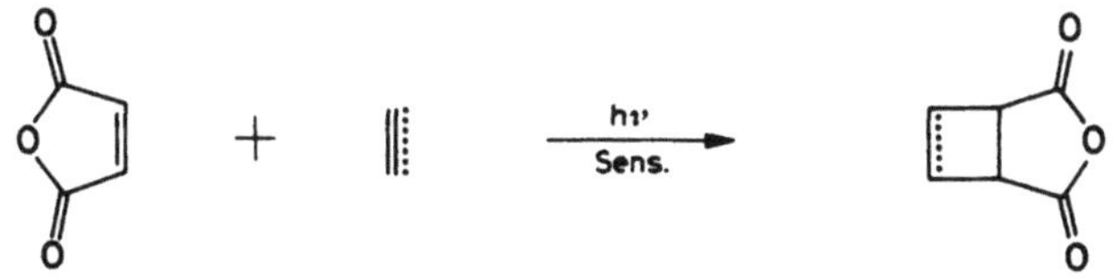

(4.70)

The unsaturated lactone in (4.71) adds to ethylene to give a precursor of the pheromone (±)-grandisol (4.71) [483].

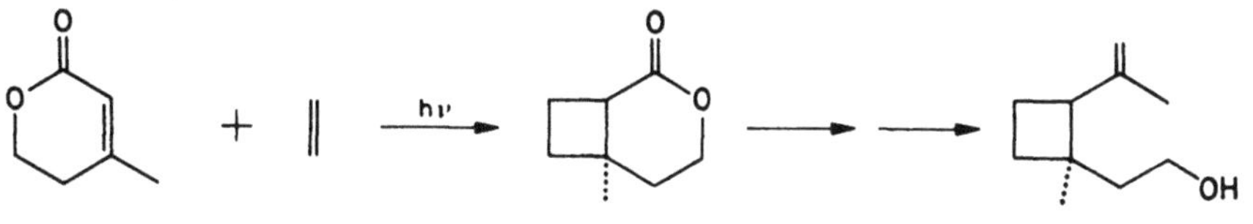

(4.71)

The cyclononadiene in (4.72) is obtained by thermal rearrangement of the photoadduct of a cyclobutene carboxylic acid ester and cyclopentene [484].

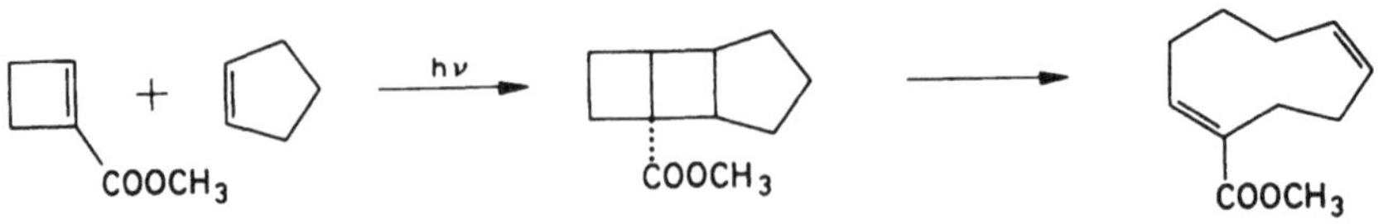

(4.72)

An elegant complementary variation of the 2-cyclohexenone synthesis in (4.63) has been achieved in using the dioxenone as photoreactant (4.73) [485].

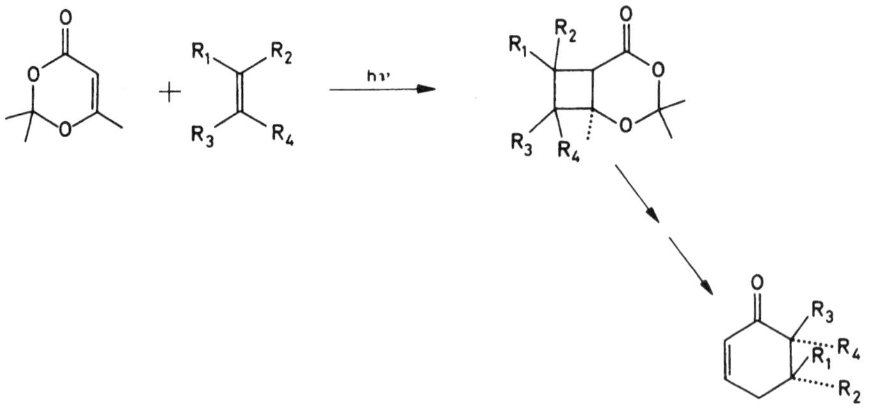

(4.73)

4.3.5 Photocycloaddition of Imines to Olefins

As mentioned in chapter 4.2.3, aliphatic imines are photochemically rather unreactive. When the C—N double bond is conjugated to an electron withdrawing group (e.g. carbonyl group), as in O-alkyl derivatives of succinimide and phthalimide, the reactivity increases and azetidines are obtained in cycloadditions to olefins [486]. A somehow similar example is the photoaddition of a 6-azauracil derivative to 2,3-dimethyl-2-butene (4.74) [487].

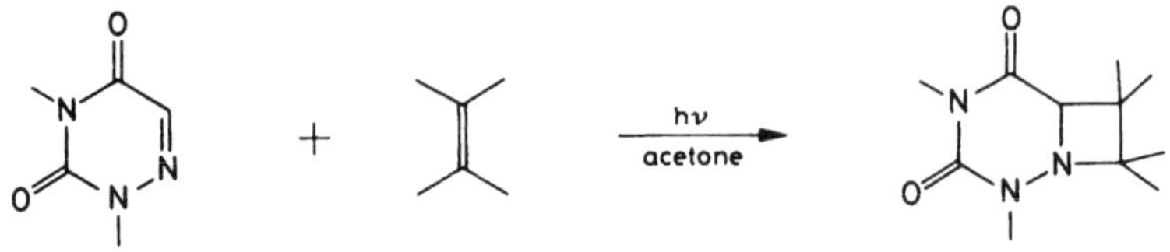

(4.74)

An interesting modification of such a reaction, wherein the reactivity of the excited chromophore is increased by protonation is outlined in the following example whereby a pyrrolenine derivative is photocyclized to a pyrrolidine derivative (4.75) [488].

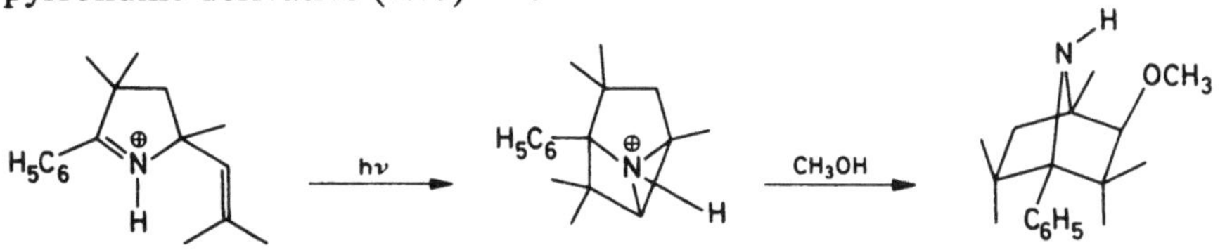

(4.75)

4.3.6 Photocycloadditions of Carbonyl Compounds to Alkenes

Oxetanes are the cycloadducts from a carbonyl compound and an olefin. This one step photochemical formation of a four membered ring heterocycle has been named the Paterno-Büchi reaction [489a, b]. Oxetanes are important synthetic intermediates as they can fragment into the carbonyl-olefin pair by which they were not formed (a so termed carbonyl-olefin metathesis). Two examples of such oxetan cracking reactions are shown below in (4.76) [490] and in (4.77) [491]; in this last example the oxetane was used as a precursor for the pheromone E-6-nonenol,

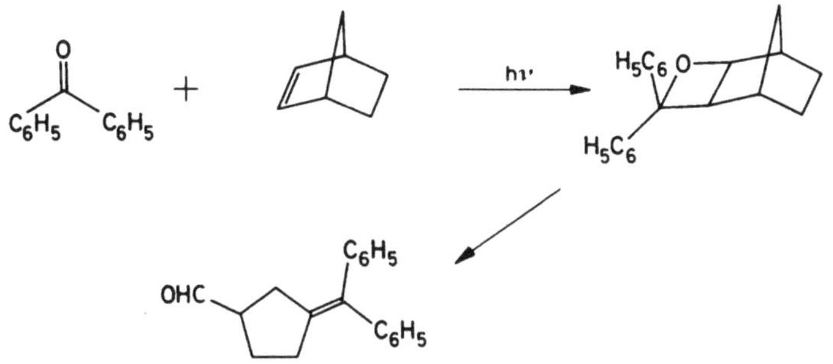

(4.76)

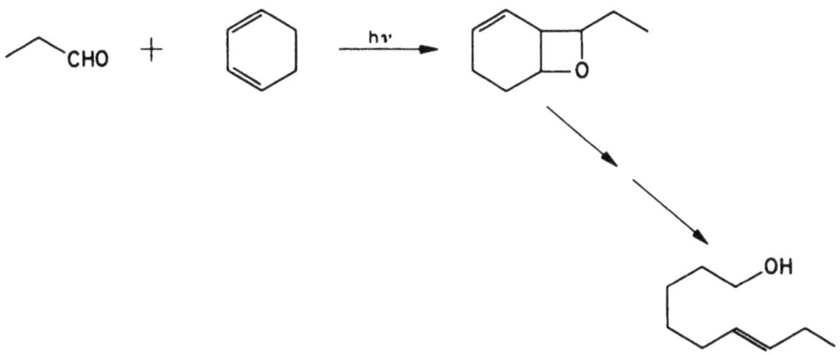

(4.77)

Polycyclic oxetanes are obtained in good yields in intramolecular carbonyl-olefin cycloadditions, in an analogous way as the corresponding alicyclic systems are formed in intramolecular enone-olefin additions. Two applications are given in (4.78) [492] and in (4.79) [493].

(4.78)

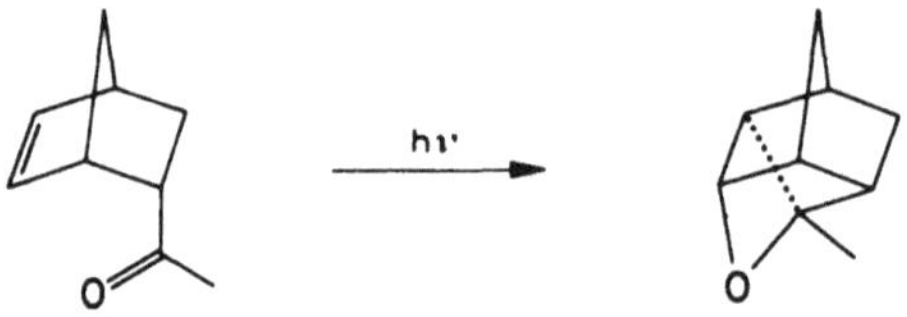

(4.79)

Benzoquinones can also serve as carbonyl compounds in this reaction. The deriving spirooxetanes can be converted to phenols (4.80) [494].

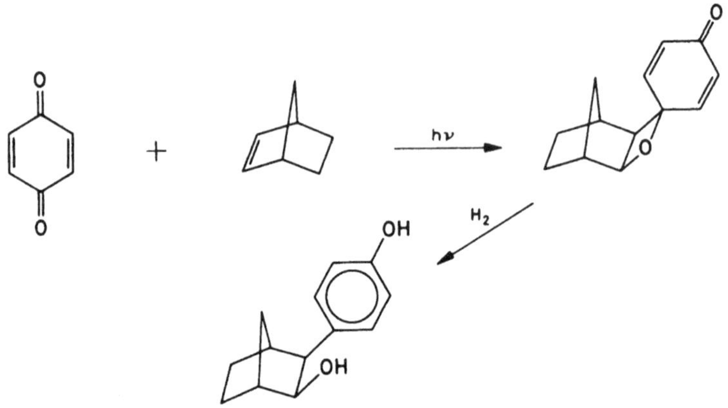

(4.80)

The mechanism of oxetane formation is similar to the one discussed for cyclobutane formation in chapter 4.3.3. The 1,4-diradicals can be efficiently trapped with molecular oxygen. The resulting 1,2,4-trioxanes are interesting synthetic intermediates (4.81) [495].

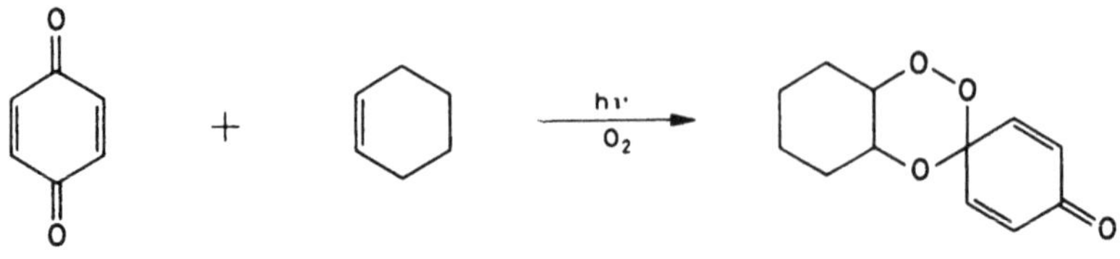

(4.81)

Imides[496] (4.82) and esters[497] (4.83) can also form oxetanes in inter- and intramolecular reactions.

(4.82)

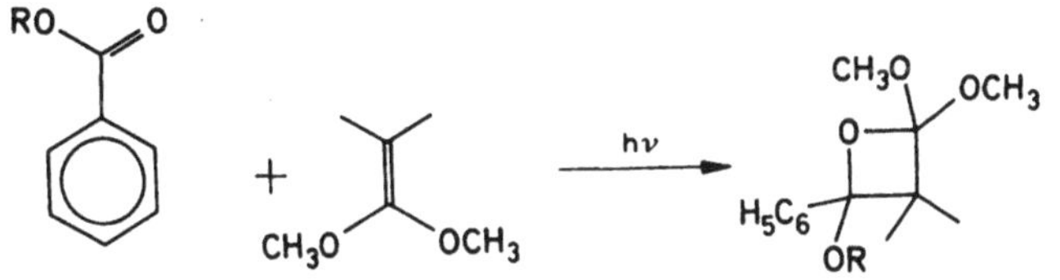

(4.83)

4.3.7 Photocycloaddition of the C—S Double Bond to Alkenes

The formation of thietanes from thiones and olefins has been less exploited for synthetic purposes than the corresponding oxetane-forming reaction. It should be remarked that thiocarbonyl compounds very often undergo efficient photoreactions from the second excited singlet state S_2 [114]. One interesting synthetic application is found in the photochemical preparation of quinolines from N-thioamides (4.84)[498]. The primary photochemical step is assumed to be the intramolecular thietane formation.

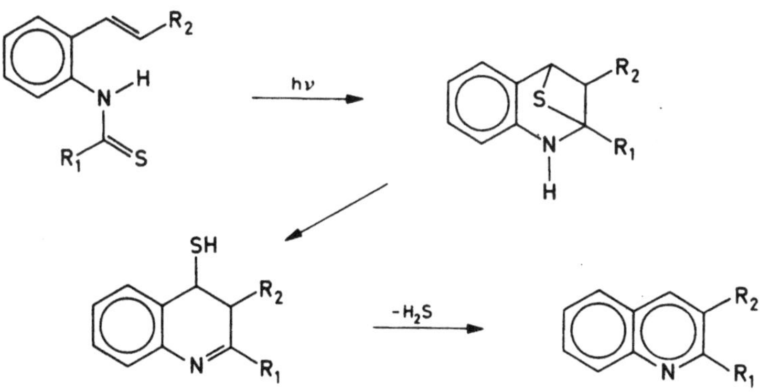

(4.84)

4.3.8 Photocycloaddition of the N—N Double Bond to Alkenes

Aliphatic azo compounds usually undergo photoextrusion of nitrogen (cf. chapter 2.3.3). Nevertheless 1,2-diazetidines are formed efficiently when there is a favourable interaction between the pi-systems of the N—N and the C—C bonds (4.85) [499].

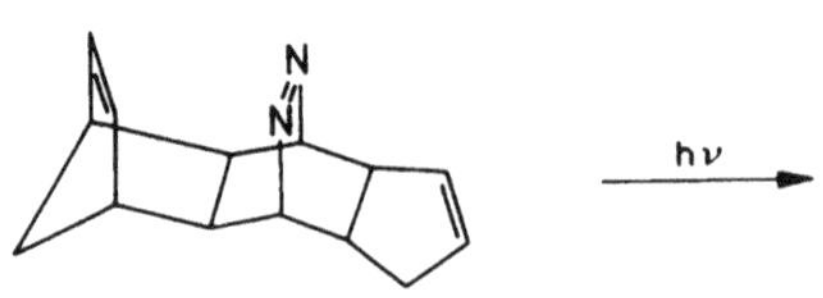

5 Photochemical Substitution Reactions in Synthetic Organic Chemistry

5.1 Aromatic Photosubstitution

Photoexcited aromatic compounds undergo substitution reactions with (non-excited) nucleophiles. The rules governing these reactions are characteristically different and often opposite to those prevailing in aromatic ground state chemistry [501a, b]. In contrast to the well known ortho/para activation in thermal aromatic substitutions, nitro groups activate the meta position in the photochemical substitution, as shown in (5.1) [502].

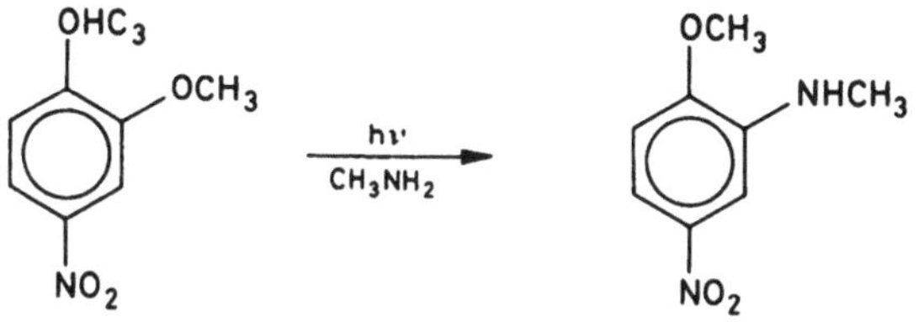

(5.1)

3-Halogenopyridines can be photohydrolized efficiently. The quantum yield of photohydrolysis is independent of the pH-value (5.2) [501a].

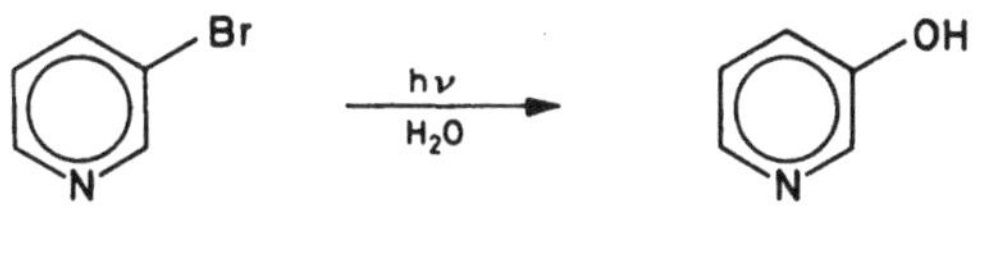

(5.2)

Aromatic compounds activated by electron donating groups undergo photosubstitution preferentially in the ortho or para position (5.3) [503].

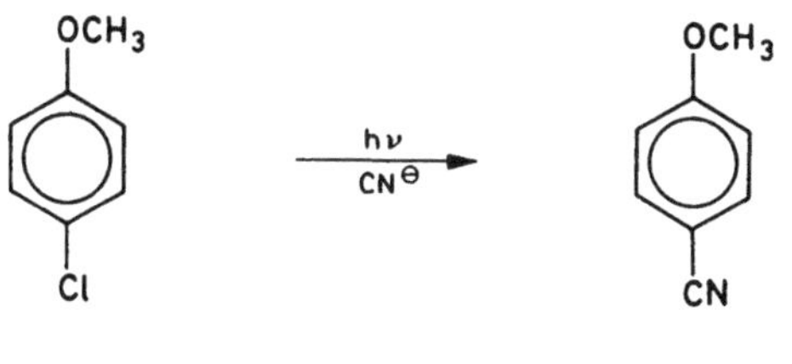

(5.3)

It seems that no general mechanistic description fits all these experiments. Some of the reactions proceed via an addition-elimination mechanism, while in others the primary step is electron transfer from the arene with formation of a radical cation. This second mechanism is then very similar to the electrochemical anodic substitution/addition sequence.

6 Photochemical Generation of Reagents for Organic Synthesis

It is the purpose of this chapter to describe photochemically generated intermediates which are used as organic reagents. These species can generally also be generated via other techniques. For carbenes or nitrenes (cf. chapters 2.1.5 and 2.1.6) the photolysis of diazo compounds, diazirines or azides is the method of choice to generate these species at low temperatures. For singlet oxygen the photochemical generation has established itself as by far the most convenient method of preparing this reactive molecule.

6.1 Singlet Oxygen

Singlet oxygen, $^1O_2$, is the first excited electronic state of molecular oxygen, lying 22.4 kcal/mol above the triplet ground state. The reaction of singlet oxygen with organic molecules is referred to as photooxydation [601-605]. Among several methods for generating $^1O_2$ in solution, as the reaction of hydrogen peroxide with sodium hypochlorite and the thermolysis of triaryl phosphite ozonides, the by far most efficient technique consists in the dye-sensitized photochemical excitation of triplet oxygen. Herein an appropriate dye (rose bengal, methylene blue or a porphyrin) is excited with visible light, populating the excited triplet state after very fast intersystem crossing. The dye acts as a sensitizer, transfering the energy to triplet oxygen to form $^1O_2$ and regenerate the dye in its ground state.

On a first, very broad, approximation singlet oxygen behaves somehow like ethylene. Three types of reactions of $^1O_2$ are usually observed and have been utilized in organic synthesis [606-608]: a) the Diels-Alder like cycloaddition to dienes (6.1); b) the ene reaction with olefins (6.2) and c) cycloaddition to activated double bonds (6.3).

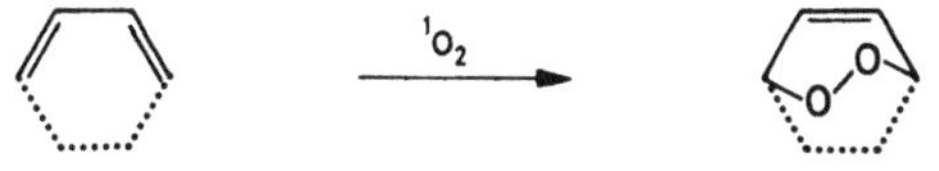

(6.1)

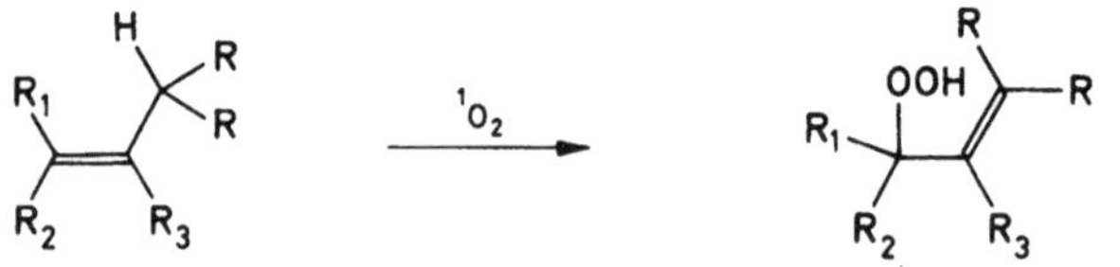

(6.2)

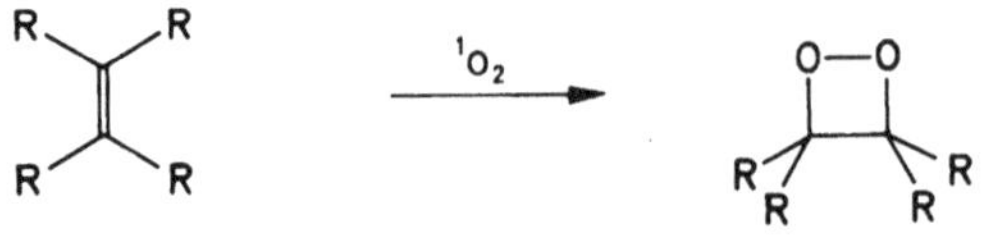

(6.3)

6.1.1 [2 + 4]Cycloadditions of $^1O_2$ to Conjugated Dienes

Singlet oxygen adds to cyclic 1,3-dienes to afford 1,4-endoperoxides [609]. This type of reaction has first been applied successfully to the synthesis of ascaridole from α-terpinene [610] (6.4) and of the vesicatory compound cantharidine [611] (6.5) some 25 years ago.

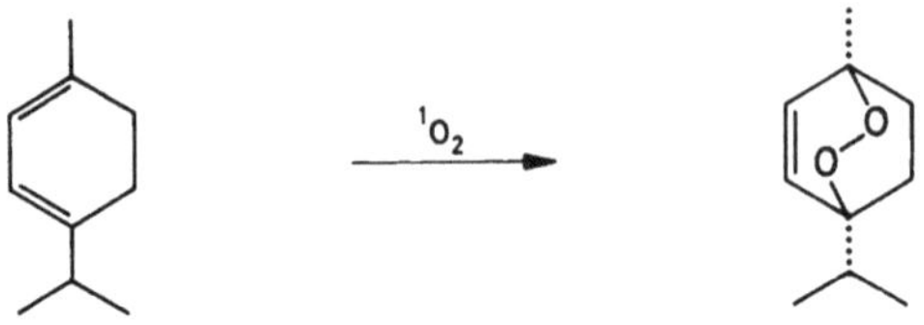

(6.4)

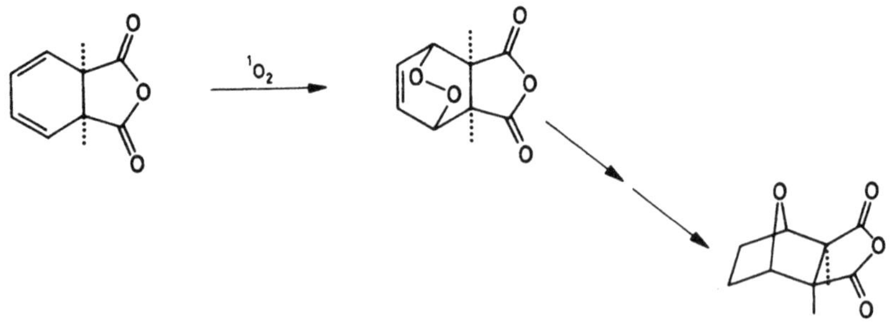

(6.5)

An interesting "umpolung" effect has been found in the alkylation of pyrroles with carbon nucleophiles via the endoperoxide (6.6) [612].

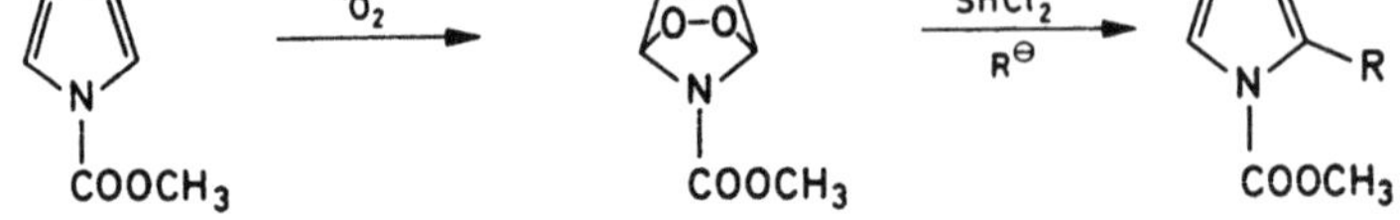

(6.6)

The corresponding endoperoxides of furans usually ring-open to afford 1,4-diketones. Such a sequence, followed by an intramolecular Diels Alder reaction was utilized in the synthesis of the *as*-indacenyl dianion (6.7) [613].

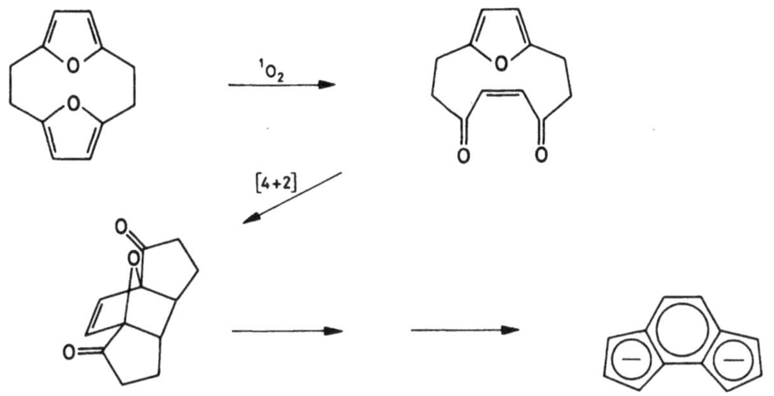

(6.7)

Endoperoxides of carbocyclic 1,3-dienes are most often transformed into either 1,4-diols, 1,3-diepoxides or 1,4-hydroxyenones. The 1,4-diol formation is illustrated in the synthesis of the sesquiterpene (±)-cybullol (6.8) [614].

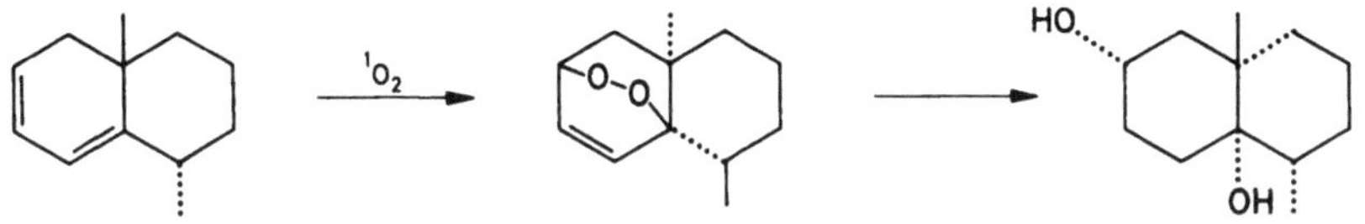

(6.8)

The diepoxide (±)-crotepoxide has been synthesized by a single [615] (6.9) or a double Diels Alder reaction (6.10) [616].

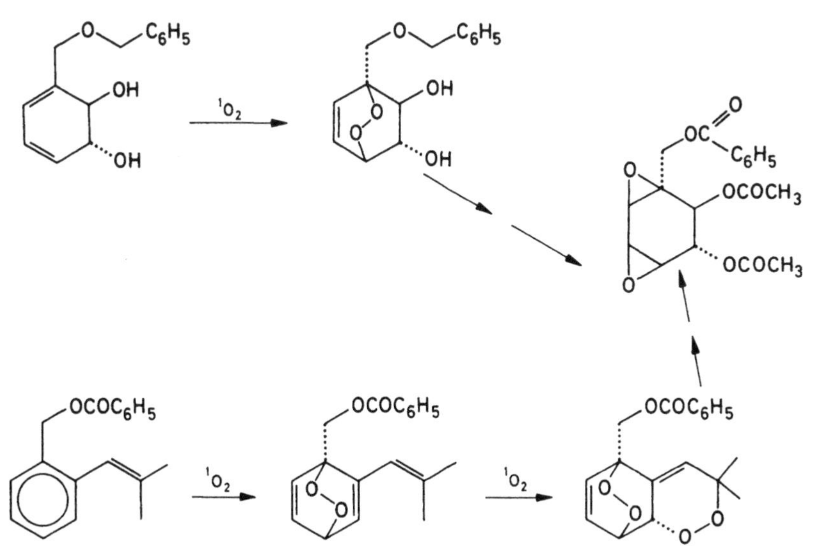

(6.9)

(6.10)

The sesquiterpene α-agarofuran was obtained via an endoperoxide → → hydroxyenone rearrangement (6.11) [617].

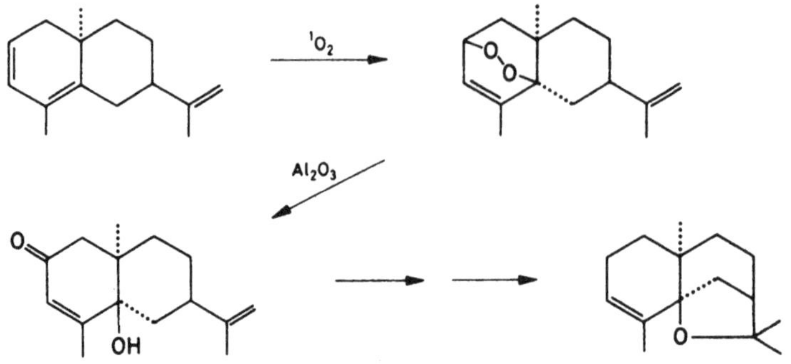

(6.11)

The furanoid terpene p. menthofuran (cf. (2.18)) has been synthesized via photooxydation of an acyclic 1,3-diene (6.12) [618].

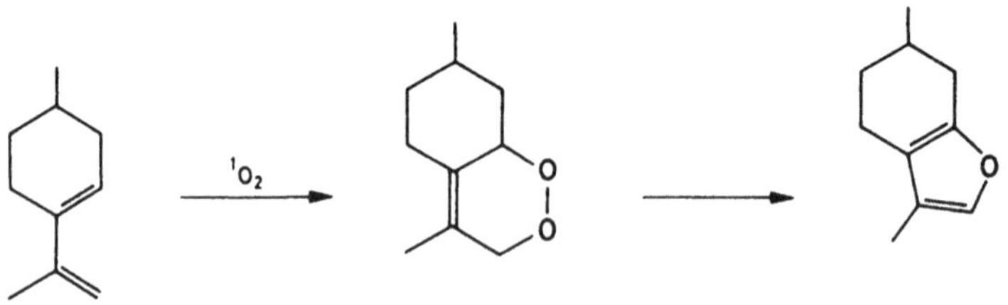

(6.12)

6.1.2 Ene-like Reactions of Singlet Oxygen

Olefins containing at least one allylic hydrogen atom react with $^1O_2$ to form an allylic hydroperoxide. The analogous formal reaction of such alkenes with ethylene is known as ene-reaction. The tricyclic lactone peroxyparthenolide has been prepared by such a reaction (6.13) [619].

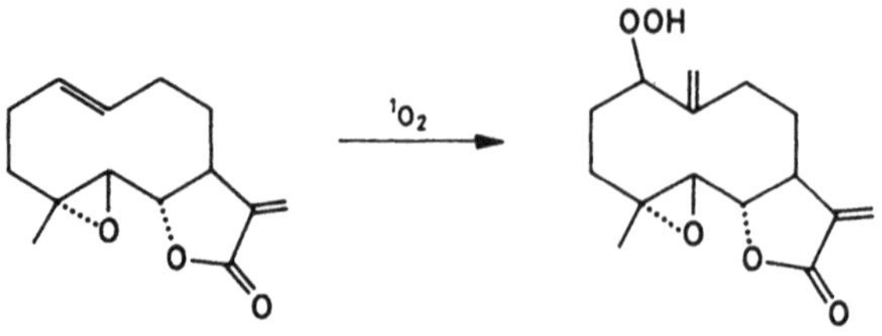

(6.13)

Usually the allylic peroxides are not the final products but they are reduced to allylic alcohols as in the synthesis of E-5-cyclodecenone (6.14) [620].

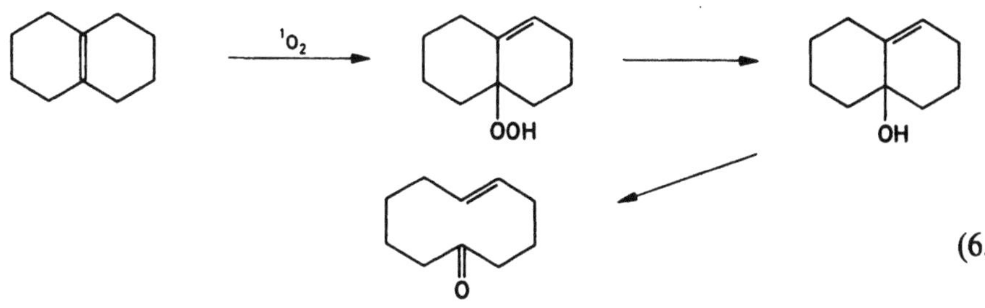

(6.14)

Alternatively the allylic hydroperoxide can undergo the so called Hock-type fragmentation [621] under acid-catalyzed hydrolysis with formation of two carbonyl compounds (or moieties). The sesquiterpene (—)-geijerone has been obtained from the triene (+)-γ-elemene by such a sequence (6.15) [622].

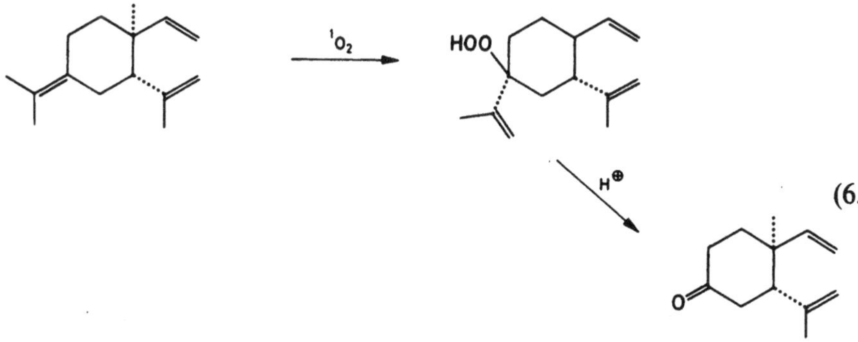

(6.15)

In special cases allylic hydroperoxides decompose directly to α,β-unsaturated ketones. The allylic hydroperoxide formed in the photooxydation of calarene gives the three types of products discussed above (6.16) [623].

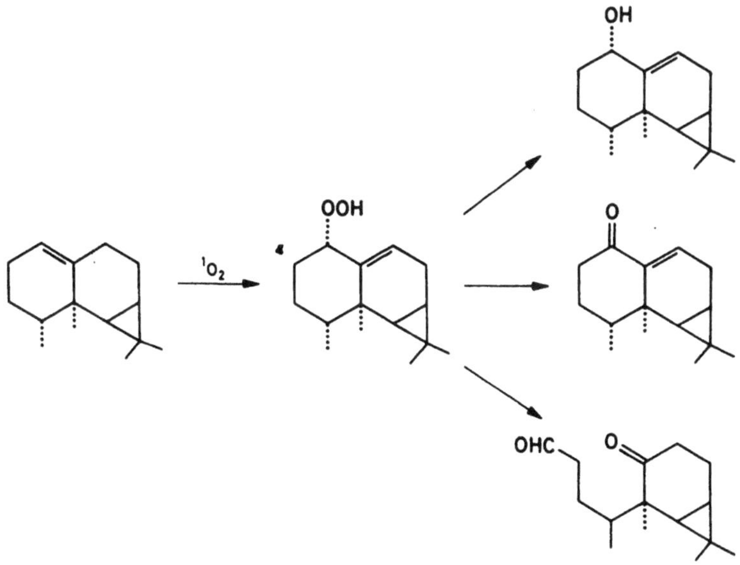

(6.16)

6.1.3 [2 + 2]Cycloadditions of $^1O_2$ to Olefins

Singlet oxygen reacts with electron rich or highly strained alkenes to form 1,2-dioxetanes. These four-membered ring peroxides decompose on warming to two carbonyl compounds (or moieties), usually with appearance of light emission (chemiluminescence). The macrocyclic bis-lactone in (6.17) [608], a musk fragrance, has been synthesized via such a sequence.

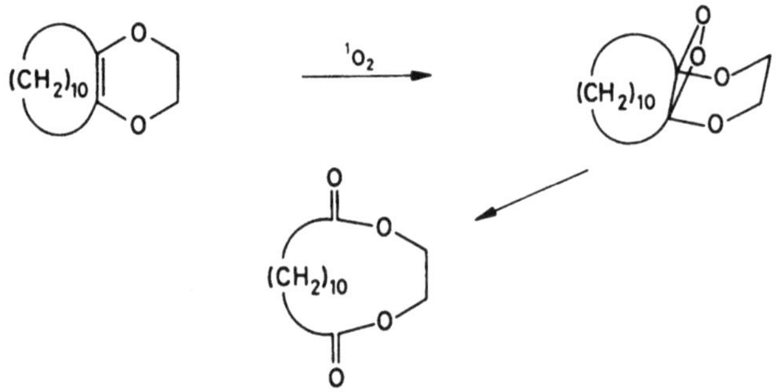

(6.17)

Enamines which cannot undergo ene-like reactions add efficiently to singlet oxygen. The intermediate dioxetane can be cleaved under mild conditions to afford a ketone (and an amide). Such a sequence has been applied in a synthesis of testosterone (6.18) [624].

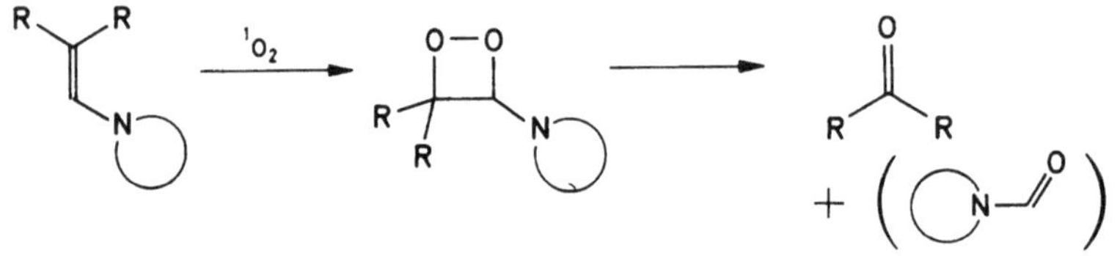

(6.18)

Use of an interesting enamine photooxidation reaction has been made of in the synthesis of β-lactams from 2-azetidine carboxylic esters (6.19) [625].

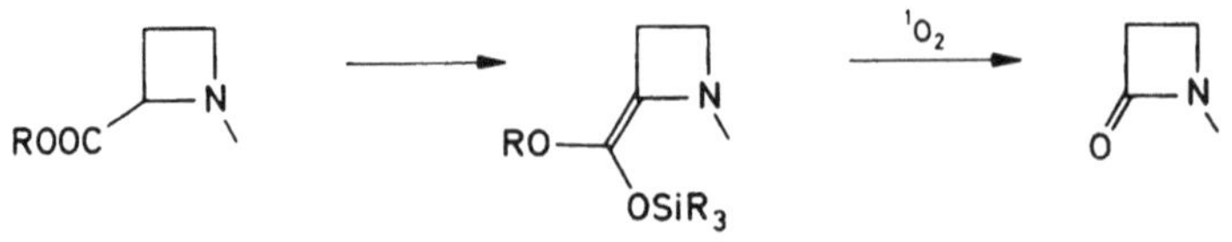

(6.19)

7 Experimental Techniques

A short survey of the most important experimental aspects will be outlined briefly although very complete descriptions of experimental methods in organic photochemical syntheses are available [120,121,701–703].

Once the UV-absorption spectrum of the compound to be irradiated — and whenever possible of the reaction product as well — is known, the main parameter to be selected is the excitation wavelength. In unsensitized — i.e. direct — irradiations the reaction product should not absorb any light as to avoid secondary photoreactions. The wavelength of the light used can be influenced by three factors: the light source, filters and the solvent.

Two types of mercury lamps are commonly used in preparative organic photochemistry; the first one, a low-pressure Hg-lamp, emits selectively at 254 nm. A quartz apparatus has to be used in combination with such a lamp as conventional glass is usually transparent only down to 285 nm. The second type is a high-pressure Hg-lamp (125–450 W) emitting light from 254 nm → 600 nm, with the main emission lines centered at 313, 366, 405, 436, 546 and 578 nm. A variety of glass filters or liquid filter solutions [120,121,701] can be used to filter out any undesired — or select any desired — wavelength region, therefore rendering expensive monochromators unnecessary for preparative experiments.

The solvent should be transparent at the wavelength used, and as unreactive as possible. Alkanes, alcohols, acetonitrile and water are transparent down to 200 nm; dichloromethane ($\lambda > 245$ nm), benzene ($\lambda > 310$ nm) and acetone ($\lambda > 325$ nm) are also often used. Isopropyl alcohol is the solvent of choice for photoreductions. Perfluorinated hydrocarbons can be used if hydrogen abstraction reactions are to be avoided, although they are rather poor solvents for polar organic molecules. In sensitized reactions the sensitizer should absorb as much of the light as possible with regard to starting material and reaction product. Acetone and benzene are conveniently used, as they posess high triplet energies ($E_T \approx 80$ kcal/mol) and they can readily be distilled off after use. Michler's ketone (4,4′-dimethylaminobenzophenone) can be extracted with acid. In using other sensitizers, separatory problems can arise when isolating the reaction product.

It is usually recomended to "degass" the solution before irradiation, although the oxygen concentration in organic solvents is rather low. This is best done by bubbling dry argon or nitrogen through the solution for one to two minutes before running the experiment.

The conventional photolysis apparatus consists of a concentrically arranged immersion well for the lamp, which is surrounded by a cooling jacket, which is itself surrounded by the reaction vessel. If this last compartment is used for the filter solution an additional external flask for the reaction mixture has to be used. There are also photochemical reactors wherein the lamps are arranged externally around the reaction flask.

As for the temperature, preparative photochemical reactions are usually run around room temperature with water as coolant for the lamp, although higher or lower temperatures can be achieved in using a thermostate or a cryostate to circulate the coolant, which obviously should be transparent for the light used. Only very few reactions in a matrix at low temperatures are used for preparative purposes [704].

Photochemical reactions are usually run in homogeneous solutions; notwithstanding it is also possible to irradiate solid compounds directly. Examples of such reactions on a preparative scale [705] as well as a discussion on crystal lattice control on photoreactions [706] are found in the literature. Finally, specific effects of a micellar environement is also being used in photochemical reactions of preparative purposes [707].

8 References

Chapter 1

101. D. R. Arnold et al.: Photochemistry, Academic Press, 1974
102. J. A. Barltrop, J. D. Coyle: Principles of Photochemistry, J. Wiley, 1975
103. D. O. Cowan, R. L. Drisko: Elements of Organic Photochemistry, Plenum Press, 1976
104. A. Cox, T. J. Kemp: Introductory Photochemistry, McGraw & Hill, 1971
105. W. M. Horspool: Aspects of Organic Photochemistry, Academic Press, 1976
106. P. Margaretha: Introduction à la Photochimie Organique in: La Bioconversion de l'Energie Solaire (ed. C. Gavach), Masson, 1981, p. 87
107. N. J. Turro: Modern Molecular Photochemistry, Benjamin, 1978
108a–j. D. Bryce-Smith: Specialist Periodical Report "Photochemistry", The Chemical Society, *1* (1970)–*10* (1979)
109. Organic Photochemistry a) *1* (1967), b) *2* (1969), c) *3* (1973), (ed. O. L. Chapman), d) *4* (1979), e) *5* (1981), (ed. A. Padwa), M. Dekker
110. Adv. Photochemistry, *1* (1963)–*12* (1980)
111. P. de Mayo: Rearrangements in Ground- and Excited States, Vol. 3, Academic Press, 1980
112. N. J. Turro, V. Ramamurthy: p. 1 in ref. 111
113. N. J. Turro: Chem. Rev. *78*, 125 (1978)
114. P. de Mayo: Acc. Chem. Res. *9*, 383 (1976)
115. A. Albini: Synthesis *1981*, 249
116. R. A. Caldwell: Acc. Chem. Res. *13*, 45 (1980)
117. F. D. Lewis: ibid. *12*, 152 (1979)
118. P. J. Wagner: Photorearrangement via Biradicals, p. 381 in ref. 111
119. W. G. Dauben, L. Salem, N. J. Turro: Acc. Chem. Res. *8*, 41 (1975)
120. A. Schönberg, G. O. Schenck, O. A. Neumüller: Preparative Organic Photochemistry, Springer, 1968
121. Houben-Weyl: Methoden der Organischen Chemie, Vol. IV/5a and IV/5b, G. Thieme, 1975
122. A. B. Holmes: Specialist Periodical Report "General and Synthetic Methods", Vol. 3 (ed. G. Pattenden), The Chemical Society, 1980, p. 329
123. H. A. J. Carless: Chem. in Britain *16*, 456 (1980)
124. R. S. Davidson: Chem. & Ind. *1978*, 180
125. R. Srinivasan, T. D. Roberts: Organic Photochemical Synthesis *1* (1971), *2* (1976), J. Wiley
126. V. N. R. Pillai: Synthesis *1980*, 1

Chapter 2

201. D. S. Weiss: p. 347 in ref. 109e
202. J. Kossanyi: Pure Appl. Chem. *51*, 181 (1979)
203. W. D. Stohrer et al.: Topics in Current Chem. *46*, 181 (1974)
204. P. Yates, R. O. Loutfy: Acc. Chem. Res. *8*, 209 (1975)
205. J. A. Barltrop, J. D. Coyle: JCS Chem. Comm. *1969*, 1081
206. R. F. Newton et al.: ibid. *1979*, 683
207. G. Ciamician, P. Siber: Chem. Ber. *41*, 1071 (1908)
208. J. Meinwald, R. A. Chapman: J. Amer. Chem. Soc. *90*, 3218 (1968)
209. P. Margaretha: Helv. Chim. Acta *59*, 2902 (1976)
210. A. F. Thomas, M. Ozainne: ibid. *59*, 1243 (1976)
211. R. G. Carlson, J. H. A. Huber, D. E. Henton: JCS Chem. Comm. *1973*, 223
212. W. C. Agosta, A. B. Smith III: J. Amer. Chem. Soc. *93*, 5513 (1971)
213. G. C. Crockett, T. H. Koch: J. Org. Chem. *42*, 2721 (1977)
214a. A. Henne, N. P. Y. Siew, K. Schaffner: Helv. Chim. Acta *62*, 1952 (1978)
214b. E. Baggiolini, H. P. Hamlow, K. Schaffner: J. Amer. Chem. Soc. *92*, 4096 (1970)
215. P. S. Engel, M. A. Schnexnayder: ibid. *97*, 145 (1975)
216. G. Quinkert: Angew. Chem. Int. Edn. *14*, 790 (1975)
217. H. Hart, D. Dickinson, W. Y. Li: Tet. Letters *1975*, 2253
218. G. Quinkert et al.: Chem. Ber. *110*, 3582 (1977)
219. G. R. Lenz: Synthesis *1978*, 489
220. N. C. Yang, G. R. Lenz: Tet. Letters *1967*, 4897
221. P. G. Sammes: in: The Chemistry of the Carbon-Halogen Bond (ed. S. Patai), J. Wiley, 1973, p. 747
222. T. Kametani et al.: JCS Chem. Comm. *1971*, 774
223. S. M. Kupchan, C. K. Kim, K. Miyano: ibid. *1976*, 91
224. R. B. Boar: p. 717 in ref. 121
225. D. H. R. Barton, J. M. Beaton: J. Amer. Chem. Soc. *84*, 1496 (1962)
226. D. H. R. Barton: Pure Appl. Chem. *16*, 1 (1968)
227. H. Dürr: p. 1158 in ref. 121
228. W. Ando: in: The Chemistry of Diazonium and Diazo Groups (ed. S. Patai), J. Wiley, 1978, p. 366
229. R. C. Woodworth, P. S. Skell: J. Amer. Chem. Soc. *81*, 3383 (1958)
230. J. J. Zupancic, P. B. Grasse, G. B. Schuster: ibid. *103*, 2423 (1981)
231. D. Schoberg: Chem. Ber. *102*, 1789 (1969)
232. W. J. Baron, M. E. Hendrick, M. Jones Jr.: J. Amer. Chem. Soc. *95*, 6286 (1973)
233. E. Wenkert et al.: ibid. *99*, 4778 (1977)
234. H. E. Zimmerman, D. H. Paskowich: ibid. *86*, 2149 (1964)
235. S. Masamune: ibid. *86*, 735 (1964)
236. C. Casagrande, L. Canonica: JCS Perkin I *1975*, 1647
237. B. T. Golding, D. R. Hall: ibid. *1975*, 1517
238. D. Becker, Z. Harel, D. Birnbaum: JCS Chem. Comm. *1975*, 377
239. T. Gibson, W. F. Erman: J. Org. Chem. *31*, 3028 (1966)
240. T. Miyashi, T. Nakayo, T. Mukai: JCS Chem. Comm. *1978*, 442
241a. E. P. Kyba, R. A. Abramovitch: J. Amer. Chem. Soc. *102*, 735 (1980)
241b. R. A. Abramovitch, E. P. Kyba: in: The Chemistry of the Azido Group (ed. S. Patai), Interscience, 1971, p. 221
242. F. D. Lewis, W. H. Saunders Jr.: Nitrenes (ed. W. Lwowski), Interscience, 1970
243a. P. Margaretha, S. Solar: Angew. Chem. Int. Ed. *11*, 1024 (1972)
243b. P. Margaretha, H. Steffen: Chimia *27*, 479 (1973)

244. J. A. Hyatt, J. S. Swenton: J. Heterocyc. Chem. *9*, 409 (1972)
245. K. Kawashima, I. Agata: J. Pharm. Soc. Japan *89*, 1426 (1969)
246. P. F. Alewood, M. Benn, R. Reinfried: Can. J. Chem. *52*, 4083 (1974)
247. G. L'abbé: Angew. Chem. Int. Edn. *14*, 775 (1975)
248. R. S. Givens: p. 227 in ref. 109e
249. G. Quinkert, C. Cimbollek, G. Buhr: Tet. Letters *1966*, 4573
250. W. G. Dauben et al.: J. Org. Chem. *41*, 887 (1976)
251. T. Kissel, R. Matusch, K. Hartke: Z. Chem. *16*, 318 (1976)
252. G. Maier et al.: Angew. Chem. Int. Edn. *17*, 520 (1978)
253. J. D. Coyle: Chem. Rev. *78*, 97 (1978)
254. W. Adam, J. C. Liu, O. Rodriguez: J. Org. Chem. *38*, 2269 (1973)
255. M. L. Kaplan, E. A. Truesdale: Tet. Letters *1976*, 3665
256. T. Otsubo, K. Kohda, S. Misumi: ibid. *1978*, 2507
257. P. S. Engel: Chem. Rev. *80*, 99 (1980)
258a. W. Adam et al.: J. Amer. Chem. Soc. *103*, 998 (1981)
258b. W. Adam, F. P. Mazenod: ibid. *102*, 7131 (1980)
259. T. J. Katz, N. J. Acton: ibid. *95*, 2738 (1973)
260. J. M. Wyrratt, L. A. Paquette: Tet. Letters *1974*, 2433
261. E. Lüddecke et al.: Tetrahedron *33*, 2677 (1977)
262. M. Schneider, O. Schuster, H. Rau: Chem. Ber. *110*, 2180 (1977)
263. R. M. Wilson, F. Geiser: J. Amer. Chem. Soc. *100*, 2225 (1978)
264. C. Dietrich-Buchecker, D. Martina, M. Franck-Neumann: J. Chem. Res. (S) *1978*,

Chapter 3

301. P. J. Kropp: p. 1 in ref. 109d
302. J. D. Coyle: Chem. Soc. Rev. *3*, 329 (1974)
303. J. Saltiel, J. L. Charlton: p. 25 in ref. 111
304a. A. Padwa: p. 261 in ref. 109d
304b. A. Padwa: Acc. Chem. Res. *12*, 310 (1979)
305. C. W. Mallory, F. B. Mallory: Org. Photochem. Synth. *1*, 55 (1971)
306. R. A. Dybans: Org. Photochem. Synth. *1*, 25 (1971)
307. P. S. Mariano, E. Krochmal Jr., A. Leone: J. Org. Chem. *42*, 1122 (1977)
308. V. Ramamurthy, R. S. H. Liu: J. Amer. Chem. Soc. *98*, 2935 (1976)
309. P. G. Gassman, E. A. Williams: Org. Photochem. Synth. *1*, 44 (1971)
310. Y. Inoue, S. Takamaku, H. Sakurai: Synthesis *1977*, 111
311. M. Nakazaki, K. Yamamoto, J. Yanagi: JCS Chem. Comm. *1977*, 346
312. H. Nozaki, M. Kurita, R. Noyori: Tet. Letters *1968*, 2025
313. Y. Shigemitsu, D. R. Arnold: JCS Chem. Comm. *1975*, 407
314. H. Meier: p. 222 in ref. 121
315. W. G. Dauben, E. L. McInnis, D. M. Michno: p. 91 in ref. 111
316. P. Courtot, R. Rumin, J. Y. Salun: Pure Appl. Chem. *49*, 317 (1977)
317. G. M. Sanders, J. Pot, E. Havinga: Fortschr. Chem. Org. Naturst. *27*, 131 (1969)
318. F. I. Sonntag, R. Srinivasan: Org. Photochem. Synth. *1*, 39 (1971)
319. D. I. Schuster, B. R. Sckolnick: ibid. p. 28
320. W. L. Dilling: Org. Photochem. Synth. *2*, 5 (1976)
321. J. S. Swenton, G. L. Snyser: J. Org. Chem. *43*, 165 (1978)
322. H. E. Zimmerman: p. 131 in ref. 111
323. S. S. Hixson, P. S. Mariano, H. E. Zimmerman: Chem. Rev. *73*, 531 (1973)

324. H. E. Zimmerman, G. Grunewald: J. Amer. Chem. Soc. *88*, 183 (1966)
325. L. A. Paquette, G. R. Krow, T. J. Barton: Org. Photochem. Synth. *1*, 67 (1971)
326. A. G. Anastassiou, E. Reichmanis, A. E. Winston: Angew. Chem. Int. Edn. *15*, 370 (1976)
327. P. S. Mariano et al.: J. Amer. Chem. Soc. *98*, 5899 (1976)
328. L. A. Paquette et al.: J. Org. Chem. *42*, 503 (1977)
329. G. Pattenden, D. Whybrow: JCS Perkin I *1981*, 1046
330a. D. Bryce-Smith, A. Gilbert: p. 349 in ref. 111
330b. D. Bryce-Smith, A. Gilbert: Tetrahedron *32*, 1309 (1976)
331. I. Haller: J. Amer. Chem. Soc. *88*, 2070 (1966)
332. D. Bryce-Smith, A. Gilbert, D. A. Robinson: Angew. Chem. Int. Edn. *10*, 745 (1971)
333. D. I. Schuster: p. 167 in ref. 111
334. K. N. Houk: Chem. Rev. *76*, 1 (1976)
335. W. G. Dauben, G. Lodder, J. Ipatschki: Top. Curr. Chem. *54*, 73 (1975)
336. K. Schaffner: Tetrahedron *32*, 641 (1976)
337. D. Bellus, D. Kearns, K. Schaffner: Helv. Chim. Acta *52*, 971 (1969)
338. D. I. Schuster, R. H. Brown, B. M. Resnick: J. Amer. Chem. Soc. *100*, 4504 (1978)
339. R. L. Cargill et al.: J. Org. Chem. *45*, 3999 (1980)
340. M. Tada et al.: Bull. Chem. Soc. Japan *49*, 1666 (1976)
341. O. L. Chapman et al.: J. Amer. Chem. Soc. *91*, 6856 (1969)
342. P. S. Engel et al.: ibid. *96*, 924 (1974)
343a. M. Demuth et al.: Helv. Chim. Acta *63*, 2434 (1980)
343b. M. Demuth et al.: ibid. p. 2440
344. K. Schaffner, M. Demuth: p. 281 in ref. 111
345. D. H. R. Barton, P. de Mayo, M. Shafik: J. Chem. Soc. *1958*, 140
346. D. Caine, A. A. Boncugnani, W. R. Pennington: J. Org. Chem. *41*, 3632 (1976)
347. D. Caine, P. F. Ingwalson: ibid. *37*, 3751 (1972)
348. D. Caine, A. S. Forbese: Tet. Letters *1977*, 3107
349. M. Nastasi, J. Streith: p. 445 in ref. 111
350. A. Padwa: p. 501 in ref. 111
351. S. T. Reid: Adv. Heterocycl. Chem. *11*, 1 (1970)
352. S. Masamune, S. Takada, R. T. Seidner: J. Amer. Chem. Soc. *91*, 7769 (1969)
353. J. L. Miesel, P. Break: Org. Photochem. Synth. *1*, 50 (1971)
354. M. Nastasi, E. S. Schilling, J. Streith: ibid. *2*, 60 (1976)
355. P. Uebelhart, P. Gilgen, H. Schmid: ibid. p. 72
356. A. Padwa, A. Ku: J. Amer. Chem. Soc. *100*, 2181 (1978)
357. H. Röttele, G. Heil, G. Schröder: Chem. Ber. *84*, 111 (1978)
358. T. Kametani, K. Fukumoto: Acc. Chem. Res. *5*, 212 (1972)
359. G. R. Lenz, Chi Min Woo: J. Heterocycl. Chem. *18*, 691 (1981)
360. P. G. Cleveland, O. L. Chapman: JCS Chem. Comm. *1967*, 1064
361. J. C. Gramain, Y. Troin, D. Vallee: ibid. *1981*, 832

Chapter 4

401. J. M. Hornback, G. S. Proehl, I. J. Starner: J. Org. Chem. *40*, 1077 (1975)
402. T. H. Koch, J. A. Oleson, J. de Niro: ibid. *40*, 14 (1975)
403. D. G. Whitten: in: Photochemistry of Heterocyclic Compounds (ed. O. Buchardt), Wiley-Interscience, 1976, p. 524
404. R. C. Cookson et al.: Tetrahedron 24, 4353 (1968)

405. K. H. Schulte-Elte et al.: Helv. Chim. Acta *54*, 1759 (1971)
406. F. D. Lewis, R. W. Johnson, R. A. Ruden: J. Amer. Chem. Soc. *94*, 4292 (1972)
407. E. C. Alexander, J. Uliana: ibid. *98*, 4324 (1976)
408. D. C. Neckers et al.: Org. Photochem. Synth. *2*, 77 (1976)
409a. L. A. Paquette, D. W. Balogh: J. Amer. Chem. Soc. *103*, 228 (1981)
409b. L. A. Paquette, D. W. Balogh: Science *211*, 575 (1981)
410. A. B. Smith III, W. C. Agosta: JCS Chem. Comm. *1971*, 343
411. J. R. Scheffer, R. E. Gayler: Org. Photochem. Synth. *2*, 80 (1976)
412. R. W. Binkley: Synth. Comm. *6*, 281 (1976)
413. P. G. Sammes: Tetrahedron *32*, 405 (1976)
414a. N. C. Yang, C. Rivas: J. Amer. Chem. Soc. *83*, 2213 (1961)
414b. G. Quinkert et al.: Nachr. Chem. Tech. Lab. *28*, 884 (1980)
415. P. Margaretha, K. Schaffner: Helv. Chim. Acta *56*, 2884 (1973)
416. P. H. Mazzocchi: p. 421 in ref. 109e
417. Y. Sato et al.: J. Amer. Chem. Soc. *98*, 2349 (1976)
418. G. Ciamician, P. Silber: Chem. Ber. *33*, 2911 (1900)
419. P. de Mayo, J. B. Stothers, W. Templeton: Can. J. Chem. *39*, 488 (1961)
420. D. I. Schuster, S. Hussain: J. Amer. Chem. Soc. *102*, 409 (1980)
421. J. Gloor, K. Schaffner: Helv. Chim. Acta *57*, 1815 (1974)
422. F. C. de Schryver et al.: Pure Appl. Chem. *49*, 237 (1977)
423. D. J. Trecker, R. S. Foote: Org. Photochem. Synth. *1*, 81 (1971)
424. R. G. Salomon et al.: J. Amer. Chem. Soc. *96*, 1145 (1974)
425. N. Nakatsuka, S. Masamune: Org. Photochem. Synth. *2*, 57 (1976)
426. P. Margaretha: Tetrahedron *27*, 6209 (1971)
427. G. Mark et al.: Tet. Letters *1973*, 237
428. J. G. Otten et al.: J. Amer. Chem. Soc. *99*, 6353 (1977)
429. J. Fritsche: J. Prakt. Chem. *101*, 337 (1866)
430. J. P. Desvergne, H. Bouas-Laurent: JCS Chem. Comm. *1978*, 403
431. T. Teitei et al.: Austr. J. Chem. *31*, 85 (1978)
432. W. L. Dilling, N. B. Tefertiller, A. B. Mitchell: Mol. Photochem. *5*, 371 (1973)
433. P. Margaretha: Helv. Chim. Acta *65*, 290 (1982)
434. R. S. H. Liu: Org. Photochem. Synth. *1*, 48 (1971)
435. W. L. Dilling, C. E. Reinecke: ibid. p. 85
436. F. I. Sonntag, R. Srinivasan: ibid. p. 97
437. T. Katsushima, R. Yamaguchi, M. Kawanisi: JCS Chem. Comm. *1975*, 692
438. R. Srinivasan: Org. Photochem. Synth. *1*, 101 (1971)
439. J. T. M. Evers, A. Mackor: Tet. Letters *1978*, 821
440. T. Wagner-Jauregg: Synthesis *1980*, 165 & 769
441. D. Bryce-Smith, A. Gilbert: Tetrahedron *33*, 2459 (1977)
442a. H. Morrison: Acc. Chem. Res. *12*, 383 (1978)
442b. H. Morrison: p. 143 in ref. 109d
443. G. L. Grunewald, J. M. Grindel: Org. Photochem. Synth. *2*, 20 (1976)
444. R. M. Bowman, C. W. Huang, J. J. McCullough: ibid. p. 28
445. W. H. F. Sasse: ibid. p. 74
446. G. Subrahmanyan: ibid. p. 99
447. P. A. Wender, J. J. Howbert: J. Amer. Chem. Soc. *103*, 688 (1981)
448. D. Bryce-Smith, B. E. Foulger, A. Gilbert: JCS Chem. Comm. *1972*, 664
449. S. W. Baldwin: p. 123 in ref. 109e
450. P. Margaretha: Chimia *29*, 203 (1975)
451. P. de Mayo: Acc. Chem. Res. *4*, 41 (1971)

452. P. Margaretha: Helv. Chim. Acta *58*, 2161 (1975); *59*, 2236 (1976); *60*, 1472 (1977)
453. P. G. Bauslaugh: Synthesis *1970*, 287
454a. S. M. Ali, T. V. Lee, S. M. Roberts: Synthesis *1977*, 155
454b. W. L. Dilling: Photochem. & Photobiol. *25*, 605 (1977)
455. F. T. Bond, H. L. Jones, L. Scerbo: Org. Photochem. Synth. *1*, 33 (1971)
456. E. W. Della, E. Cotsaris, P. T. Hine: J. Amer. Chem. Soc. *103*, 4131 (1981)
457. S. Wolff, S. Ayral-Kaloustian, W. C. Agosta: J. Org. Chem. *41*, 2947 (1976)
458. M. Mellor, D. A. Otieno, G. Pattenden: JCS Chem. Comm. *1978*, 138
459. W. Fröstl, P. Margaretha: Helv. Chim. Acta *59*, 2244 (1976)
460a. V. Georgian, M. Saltzman: Tet. Letters *1972*, 4315
460b. P. Gund, T. M. Gund: J. Amer. Chem. Soc. *103*, 4458 (1981)
461. A. M. Birch, G. Pattenden: JCS Chem. Comm. *1980*, 1195
462. M. Pirrung: J. Amer. Chem. Soc. *103*, 82 (1981)
463. J. S. H. Kueh, M. Mellor, G. Pattenden: JCS Perkin I *1981*, 1052
464. D. Becker, D. Birnbaum: J. Org. Chem. *45*, 570 (1980)
465. D. Termont, D. de Keukeleire, M. Vandewalle: JCS Perkin I *1977*, 2349
466. P. A. Wender, J. C. Lechleiter: J. Amer. Chem. Soc. *100*, 4321 (1978)
467. P. A. Wender, J. C. Hubbs: J. Org. Chem. *45*, 365 (1980)
468. P. A. Wender, L. J. Letendere: ibid. *45*, 367 (1980)
469a. K. B. Wiberg et al.: J. Amer. Chem. Soc. *102*, 7467 (1980)
469b. S. Wolff, W. C. Agosta: JCS Chem. Comm. *1981*, 118
470. J. R. Williams, T. J. Caggiano: Synthesis *1980*, 1024
471. H. Takeshita et al.: Bull. Chem. Soc. Japan *53*, 3641 (1980)
472. A. B. Smith III, P. J. Jerris: J. Amer. Chem. Soc. *103*, 194 (1981)
473. S. W. Baldwin, M. T. Crimmins: Tet. Letters *1978*, 4197
474. S. W. Baldwin, J. M. Wilkinson: ibid. *1979*, 2657
475. S. W. Baldwin, M. T. Crimmins: J. Amer. Chem. Soc. *102*, 1198 (1980)
476a. K. Wiesner: Tetrahedron *31*, 1658 (1975)
476b. G. Marini-Bettolo et al.: ibid. *36*, 719 (1980)
477. R. L. Cargill: J. Org. Chem. *45*, 3929 (1980)
478a. K. Wiesner et al.: Can. J. Chem. *47*, 433 (1969)
478b. T. Y. R. Tsai et al.: Heterocycles *7*, 217 (1977)
479a. G. R. Lenz: Rev. Chem. Interm. *4*, 369 (1981)
479b. G. R. Lenz: J. Org. Chem. *44*, 1382 (1979)
480. B. A. Pearlman: J. Amer. Chem. Soc. *101*, 6404 (1979)
481. W. R. Baker, P. D. Senter, R. M. Coates: JCS Chem. Comm. *1980*, 1011
482a. J. J. Bloomfield, D. C. Owsley: Org. Photochem. Synth. *2*, 32 (1976)
482b. J. J. Bloomfield, D. C. Owsley: ibid. p. 36
483. H. Kosugi et al.: Bull. Chem. Soc. Japan *49*, 520 (1976)
484. P. A. Wender, J. C. Lechleiter: J. Amer. Chem. Soc. *99*, 267 (1977)
485. S. W. Baldwin, J. M. Wilkinson: ibid. *102*, 3634 (1980)
486. T. H. Koch: Topics in Current Chem. *75*, 65 (1978)
487. J. S. Swenton, J. A. Hyatt: J. Amer. Chem. Soc. *96*, 4879 (1974)
488. P. S. Mariano, J. L. Stavinoha, R. Swanson: ibid. *99*, 6781 (1977)
489a. G. Jones II: p. 1 in ref. 109e
489b. D. R. Arnold: Adv. Photochem. *6*, 301 (1968)
490. G. Jones II, S. B. Schwartz, M. T. Marton: JCS Chem. Comm. *1973*, 372
491. G. Jones II, M. A. Acquadro, M. A. Carmody: ibid. *1975*, 206
492. J. Meinwald, A. T. Hamner: ibid. *1969*, 1302
493. R. R. Sauers, W. Schinski, B. Sickles: Org. Photochem. Synth. *1*, 76 (1971)
494. E. A. Fehnel, F. C. Brockaw: J. Org. Chem. *45*, 578 (1980)

495. R. M. Wilson, S. W. Wunderly: J. Amer. Chem. Soc. *96*, 7350 (1974)
496. K. Maruyama, Y. Kubo: J. Org. Chem. *42*, 3215 (1977)
497. T. S. Cantrell: JCS Chem. Comm. *1973*, 468
498. P. de Mayo, L. K. Sydnes, G. Wenska: ibid. *1979*, 499
499. W. Berning, S. Hünig: Angew. Chem. Int. Edn. *16*, 777 (1977)

Chapter 5

501 a. E. Havinga, J. Cornelisse: Pure Appl. Chem. *47*, 1 (1976)
501 b. J. Cornelisse, G. P. de Gunst, E. Havinga: Chem. Rev. *75*, 353 (1975)
502. C. Frater, J. Cornelisse: Org. Photochem. Synth. *1*, 74 (1971)
503. J. den Heijer: ref. 6 in ref. 501 a

Chapter 6

601. H. H. Wasserman, R. W. Murray: Singlet Oxygen, Academic Press, 1979
602. A. P. Schaap: Singlet Molecular Oxygen, Dowden, Hutchinson and Ross, 1976
603. R. W. Denny, A. Nickon: Org. Reactions *20*, 133 (1973)
604. D. R. Kearns: Chem. Rev. *71*, 395 (1971)
605. C. S. Foote: Acc. Chem. Res. *1*, 104 (1968)
606. H. H. Wasserman, J. L. Ives: Tetrahedron *37*, 1825 (1981)
607. M. Matsumoto, K. Kondo: J. Synth. Org. Chem. Japan *35*, 188 (1977)
608. G. Ohloff: Pure Appl. Chem. *43*, 481 (1975)
609. K. Gollnick, G. O. Schenck: in: 1,4-Cycloaddition Reactions (ed. T. Hamer), Academic Press, 1967, p. 255
610. G. O. Schenck, K. Ziegler: Naturwiss. *32*, 157 (1954)
611. G. O. Schenck: Angew. Chem. *69*, 579 (1957)
612. M. Natsume, H. Muratake: Tet. Letters *1979*, 3477
613. T. J. Katz, V. Balogh, J. Schulman: J. Amer. Chem. Soc. *90*, 734 (1968)
614. W. A. Ayer, L. M. Browne, S. Fung: Can. J. Chem. *54*, 3276 (1976)
615. M. R. Demuth, P. E. Garrett, J. D. White: J. Amer. Chem. Soc. *98*, 634 (1976)
616. M. Matsumoto, S. Dobashi, K. Kuroda: Tet. Letters *1977*, 3361
617. H. C. Barrett, G. Büchi: J. Amer. Chem. Soc. *89*, 5665 (1967)
618. B. Harirchian, P. D. Magnus: Synth. Comm. *7*, 119 (1977)
619. F. S. El Feraly et al.: Tet. Letters *1977*, 1973
620. P. S. Wharton, G. A. Hiegel, R. V. Coombs: J. Org. Chem. *28*, 3217 (1963)
621. H. Hock, H. Kropf: Angew. Chem. *69*, 313 (1957)
622. A. F. Thomas: Helv. Chim. Acta *55*, 2429 (1972)
623. H. Takeshita, I. Shimooda, T. Hatsui: Bull. Chem. Soc. Japan *53*, 3721 (1980)
624. J. E. Huber: Tet. Letters *1968*, 3271
625. H. H. Wasserman, B. H. Lipshutz, J. S. Wu: Heterocycles *7*, 321 (1977)

Chapter 7

701. S. L. Murov: Handbook of Photochemistry, M. Dekker, 1973
702. R. Srinivasan: p. 1 in ref. 125a
703. W. M. Hardham: in: Creation and Detection of the Excited State (ed. A. A. Lamola), Dekker, 1971, p. 615
704. I. R. Dunkin: Chem. Soc. Rev. *1980*, 1
705. D. Döpp: Org. Photochem. Synth. *2*, 43 (1976)
706. J. R. Scheffer: Acc. Chem. Res. *13*, 283 (1980)
707. P. de Mayo, L. K. Sydnes: JCS Chem. Comm. *1980*, 994

9 Author Index

10 Product Index

Bei Fragen zur Produktsicherheit wenden Sie sich bitte an:
If you have any questions regarding product safety,
please contact:

Walter de Gruyter GmbH
Genthiner Straße 13
10785 Berlin
productsafety@degruyterbrill.com